FREE Test Taking Tips DVD Offer

To help us better serve you, we have developed a Test Taking Tips DVD that we would like to give you for FREE. **This DVD covers world-class test taking tips that you can use to be even more successful when you are taking your test.**

All that we ask is that you email us your feedback about your study guide. Please let us know what you thought about it – whether that is good, bad or indifferent.

To get your **FREE Test Taking Tips DVD**, email freedvd@studyguideteam.com with "FREE DVD" in the subject line and the following information in the body of the email:

a. The title of your study guide.

b. Your product rating on a scale of 1-5, with 5 being the highest rating.

c. Your feedback about the study guide. What did you think of it?

d. Your full name and shipping address to send your free DVD.

If you have any questions or concerns, please don't hesitate to contact us at freedvd@studyguideteam.com.

Thanks again!

ACT Science Tutor

ACT Science Prep Book 2020 and 2021 with 3 Practice Tests

[Includes Detailed Answer Explanations]

Test Prep Books

Table of Contents

Quick Overview

As you draw closer to taking your exam, effective preparation becomes more and more important. Thankfully, you have this study guide to help you get ready. Use this guide to help keep your studying on track and refer to it often.

This study guide contains several key sections that will help you be successful on your exam. The guide contains tips for what you should do the night before and the day of the test. Also included are test-taking tips. Knowing the right information is not always enough. Many well-prepared test takers struggle with exams. These tips will help equip you to accurately read, assess, and answer test questions.

A large part of the guide is devoted to showing you what content to expect on the exam and to helping you better understand that content. In this guide are practice test questions so that you can see how well you have grasped the content. Then, answer explanations are provided so that you can understand why you missed certain questions.

Don't try to cram the night before you take your exam. This is not a wise strategy for a few reasons. First, your retention of the information will be low. Your time would be better used by reviewing information you already know rather than trying to learn a lot of new information. Second, you will likely become stressed as you try to gain a large amount of knowledge in a short amount of time. Third, you will be depriving yourself of sleep. So be sure to go to bed at a reasonable time the night before. Being well-rested helps you focus and remain calm.

Be sure to eat a substantial breakfast the morning of the exam. If you are taking the exam in the afternoon, be sure to have a good lunch as well. Being hungry is distracting and can make it difficult to focus. You have hopefully spent lots of time preparing for the exam. Don't let an empty stomach get in the way of success!

When travelling to the testing center, leave earlier than needed. That way, you have a buffer in case you experience any delays. This will help you remain calm and will keep you from missing your appointment time at the testing center.

Be sure to pace yourself during the exam. Don't try to rush through the exam. There is no need to risk performing poorly on the exam just so you can leave the testing center early. Allow yourself to use all of the allotted time if needed.

Remain positive while taking the exam even if you feel like you are performing poorly. Thinking about the content you should have mastered will not help you perform better on the exam.

Once the exam is complete, take some time to relax. Even if you feel that you need to take the exam again, you will be well served by some down time before you begin studying again. It's often easier to convince yourself to study if you know that it will come with a reward!

Test-Taking Strategies

1. Predicting the Answer

When you feel confident in your preparation for a multiple-choice test, try predicting the answer before reading the answer choices. This is especially useful on questions that test objective factual knowledge. By predicting the answer before reading the available choices, you eliminate the possibility that you will be distracted or led astray by an incorrect answer choice. You will feel more confident in your selection if you read the question, predict the answer, and then find your prediction among the answer choices. After using this strategy, be sure to still read all of the answer choices carefully and completely. If you feel unprepared, you should not attempt to predict the answers. This would be a waste of time and an opportunity for your mind to wander in the wrong direction.

2. Reading the Whole Question

Too often, test takers scan a multiple-choice question, recognize a few familiar words, and immediately jump to the answer choices. Test authors are aware of this common impatience, and they will sometimes prey upon it. For instance, a test author might subtly turn the question into a negative, or he or she might redirect the focus of the question right at the end. The only way to avoid falling into these traps is to read the entirety of the question carefully before reading the answer choices.

3. Looking for Wrong Answers

Long and complicated multiple-choice questions can be intimidating. One way to simplify a difficult multiple-choice question is to eliminate all of the answer choices that are clearly wrong. In most sets of answers, there will be at least one selection that can be dismissed right away. If the test is administered on paper, the test taker could draw a line through it to indicate that it may be ignored; otherwise, the test taker will have to perform this operation mentally or on scratch paper. In either case, once the obviously incorrect answers have been eliminated, the remaining choices may be considered. Sometimes identifying the clearly wrong answers will give the test taker some information about the correct answer. For instance, if one of the remaining answer choices is a direct opposite of one of the eliminated answer choices, it may well be the correct answer. The opposite of obviously wrong is obviously right! Of course, this is not always the case. Some answers are obviously incorrect simply because they are irrelevant to the question being asked. Still, identifying and eliminating some incorrect answer choices is a good way to simplify a multiple-choice question.

4. Don't Overanalyze

Anxious test takers often overanalyze questions. When you are nervous, your brain will often run wild, causing you to make associations and discover clues that don't actually exist. If you feel that this may be a problem for you, do whatever you can to slow down during the test. Try taking a deep breath or counting to ten. As you read and consider the question, restrict yourself to the particular words used by the author. Avoid thought tangents about what the author *really* meant, or what he or she was *trying* to say. The only things that matter on a multiple-choice test are the words that are actually in the question. You must avoid reading too much into a multiple-choice question, or supposing that the writer meant something other than what he or she wrote.

5. No Need for Panic

It is wise to learn as many strategies as possible before taking a multiple-choice test, but it is likely that you will come across a few questions for which you simply don't know the answer. In this situation, avoid panicking. Because most multiple-choice tests include dozens of questions, the relative value of a single wrong answer is small. As much as possible, you should compartmentalize each question on a multiple-choice test. In other words, you should not allow your feelings about one question to affect your success on the others. When you find a question that you either don't understand or don't know how to answer, just take a deep breath and do your best. Read the entire question slowly and carefully. Try rephrasing the question a couple of different ways. Then, read all of the answer choices carefully. After eliminating obviously wrong answers, make a selection and move on to the next question.

6. Confusing Answer Choices

When working on a difficult multiple-choice question, there may be a tendency to focus on the answer choices that are the easiest to understand. Many people, whether consciously or not, gravitate to the answer choices that require the least concentration, knowledge, and memory. This is a mistake. When you come across an answer choice that is confusing, you should give it extra attention. A question might be confusing because you do not know the subject matter to which it refers. If this is the case, don't eliminate the answer before you have affirmatively settled on another. When you come across an answer choice of this type, set it aside as you look at the remaining choices. If you can confidently assert that one of the other choices is correct, you can leave the confusing answer aside. Otherwise, you will need to take a moment to try to better understand the confusing answer choice. Rephrasing is one way to tease out the sense of a confusing answer choice.

7. Your First Instinct

Many people struggle with multiple-choice tests because they overthink the questions. If you have studied sufficiently for the test, you should be prepared to trust your first instinct once you have carefully and completely read the question and all of the answer choices. There is a great deal of research suggesting that the mind can come to the correct conclusion very quickly once it has obtained all of the relevant information. At times, it may seem to you as if your intuition is working faster even than your reasoning mind. This may in fact be true. The knowledge you obtain while studying may be retrieved from your subconscious before you have a chance to work out the associations that support it. Verify your instinct by working out the reasons that it should be trusted.

8. Key Words

Many test takers struggle with multiple-choice questions because they have poor reading comprehension skills. Quickly reading and understanding a multiple-choice question requires a mixture of skill and experience. To help with this, try jotting down a few key words and phrases on a piece of scrap paper. Doing this concentrates the process of reading and forces the mind to weigh the relative importance of the question's parts. In selecting words and phrases to write down, the test taker thinks about the question more deeply and carefully. This is especially true for multiple-choice questions that are preceded by a long prompt.

9. Subtle Negatives

One of the oldest tricks in the multiple-choice test writer's book is to subtly reverse the meaning of a question with a word like *not* or *except*. If you are not paying attention to each word in the question, you can easily be led astray by this trick. For instance, a common question format is, "Which of the following is...?" Obviously, if the question instead is, "Which of the following is not...?," then the answer will be quite different. Even worse, the test makers are aware of the potential for this mistake and will include one answer choice that would be correct if the question were not negated or reversed. A test taker who misses the reversal will find what he or she believes to be a correct answer and will be so confident that he or she will fail to reread the question and discover the original error. The only way to avoid this is to practice a wide variety of multiple-choice questions and to pay close attention to each and every word.

10. Reading Every Answer Choice

It may seem obvious, but you should always read every one of the answer choices! Too many test takers fall into the habit of scanning the question and assuming that they understand the question because they recognize a few key words. From there, they pick the first answer choice that answers the question they believe they have read. Test takers who read all of the answer choices might discover that one of the latter answer choices is actually *more* correct. Moreover, reading all of the answer choices can remind you of facts related to the question that can help you arrive at the correct answer. Sometimes, a misstatement or incorrect detail in one of the latter answer choices will trigger your memory of the subject and will enable you to find the right answer. Failing to read all of the answer choices is like not reading all of the items on a restaurant menu: you might miss out on the perfect choice.

11. Spot the Hedges

One of the keys to success on multiple-choice tests is paying close attention to every word. This is never truer than with words like almost, most, some, and sometimes. These words are called "hedges" because they indicate that a statement is not totally true or not true in every place and time. An absolute statement will contain no hedges, but in many subjects, the answers are not always straightforward or absolute. There are always exceptions to the rules in these subjects. For this reason, you should favor those multiple-choice questions that contain hedging language. The presence of qualifying words indicates that the author is taking special care with his or her words, which is certainly important when composing the right answer. After all, there are many ways to be wrong, but there is only one way to be right! For this reason, it is wise to avoid answers that are absolute when taking a multiple-choice test. An absolute answer is one that says things are either all one way or all another. They often include words like *every*, *always*, *best*, and *never*. If you are taking a multiple-choice test in a subject that doesn't lend itself to absolute answers, be on your guard if you see any of these words.

12. Long Answers

In many subject areas, the answers are not simple. As already mentioned, the right answer often requires hedges. Another common feature of the answers to a complex or subjective question are qualifying clauses, which are groups of words that subtly modify the meaning of the sentence. If the question or answer choice describes a rule to which there are exceptions or the subject matter is complicated, ambiguous, or confusing, the correct answer will require many words in order to be expressed clearly and accurately. In essence, you should not be deterred by answer choices that seem excessively long. Oftentimes, the author of the text will not be able to write the correct answer without offering some qualifications and modifications. Your job is to read the answer choices thoroughly and completely and to select the one that most accurately and precisely answers the question.

13. Restating to Understand

Sometimes, a question on a multiple-choice test is difficult not because of what it asks but because of how it is written. If this is the case, restate the question or answer choice in different words. This process serves a couple of important purposes. First, it forces you to concentrate on the core of the question. In order to rephrase the question accurately, you have to understand it well. Rephrasing the question will concentrate your mind on the key words and ideas. Second, it will present the information to your mind in a fresh way. This process may trigger your memory and render some useful scrap of information picked up while studying.

14. True Statements

Sometimes an answer choice will be true in itself, but it does not answer the question. This is one of the main reasons why it is essential to read the question carefully and completely before proceeding to the answer choices. Too often, test takers skip ahead to the answer choices and look for true statements. Having found one of these, they are content to select it without reference to the question above. Obviously, this provides an easy way for test makers to play tricks. The savvy test taker will always read the entire question before turning to the answer choices. Then, having settled on a correct answer choice, he or she will refer to the original question and ensure that the selected answer is relevant. The mistake of choosing a correct-but-irrelevant answer choice is especially common on questions related to specific pieces of objective knowledge. A prepared test taker will have a wealth of factual knowledge at his or her disposal, and should not be careless in its application.

15. No Patterns

One of the more dangerous ideas that circulates about multiple-choice tests is that the correct answers tend to fall into patterns. These erroneous ideas range from a belief that B and C are the most common right answers, to the idea that an unprepared test-taker should answer "A-B-A-C-A-D-A-B-A." It cannot be emphasized enough that pattern-seeking of this type is exactly the WRONG way to approach a multiple-choice test. To begin with, it is highly unlikely that the test maker will plot the correct answers according to some predetermined pattern. The questions are scrambled and delivered in a random order. Furthermore, even if the test maker was following a pattern in the assignation of correct answers, there is no reason why the test taker would know which pattern he or she was using. Any attempt to discern a pattern in the answer choices is a waste of time and a distraction from the real work of taking the test. A test taker would be much better served by extra preparation before the test than by reliance on a pattern in the answers.

FREE DVD OFFER

Don't forget that doing well on your exam includes both understanding the test content and understanding how to use what you know to do well on the test. We offer a completely FREE Test Taking Tips DVD that covers world class test taking tips that you can use to be even more successful when you are taking your test.

All that we ask is that you email us your feedback about your study guide. To get your **FREE Test Taking Tips DVD**, email freedvd@studyguideteam.com with "FREE DVD" in the subject line and the following information in the body of the email:

- The title of your study guide.
- Your product rating on a scale of 1-5, with 5 being the highest rating.
- Your feedback about the study guide. What did you think of it?
- Your full name and shipping address to send your free DVD.

Introduction to the ACT

Function of the Test

The ACT is one of two national standardized college entrance examinations (the SAT being the other). Most prospective college students take the ACT or the SAT, and it is increasingly common for students to take both. All four-year colleges and universities in the United States accept the ACT for admissions purposes, and some require it. Some of those schools also use ACT subject scores for placement purposes. Sixteen states also require all high school juniors to take the ACT as part of the states' school evaluation efforts.

The vast majority of people taking the ACT are high school juniors and seniors who intend to apply to college. Traditionally, the SAT was more commonly taken than the ACT, particularly among students on the East and West coasts. However, the popularity of the ACT has grown dramatically in recent years and is now commonly taken by students in all fifty states. In fact, starting in 2013, more test takers took the ACT than the SAT. In 2015, 1.92 million students took the ACT. About 28 percent of 2015 high school graduates taking the ACT met the test's college-readiness benchmarks in all four subjects, while 31 percent met none of the benchmarks.

Test Administration

The ACT is offered on six dates throughout the year in the U.S. and Canada, and on five of those same dates in other countries. The registration fee includes score reports for four colleges, with additional reports available for purchase. There is a separate registration fee for the optional writing section.

On test dates, the ACT is administered at test centers throughout the world. The test centers are usually high schools or colleges, with several locations usually available in significant population centers.

Test takers can retake the ACT as frequently as the test is offered, up to a maximum of twelve times; although, individual colleges may have limits on how many retakes they will consider. Scores from the various sections cannot be combined from different sessions. The ACT does provide reasonable accommodations to test takers with professionally-documented disabilities.

Test Format

The ACT consists of 215 multiple-choice questions in four subject areas (English, mathematics, reading, and science) and takes about three hours and thirty minutes to complete. It also has an optional writing test, which takes an additional forty minutes.

The English section is 45 minutes long and contains 75 questions on usage, language mechanics, and rhetorical skills. The Mathematics section is 60 minutes long and contains 60 questions on algebra, geometry, and elementary trigonometry. Calculators that meet the ACT's calculator policy are permitted on the Mathematics section. The Reading section is 35 minutes long and contains four written passages with ten questions per passage. The Science section is 35 minutes long and contains 40 questions.

The Writing section is forty minutes long and is always given at the end so that test takers not wishing to take it may leave after completing the other four sections. This section consists of one essay in which students must analyze three different perspectives on a broad social issue. Although the Writing section is optional, some colleges do require it.

Section	Length	Questions
English	45 minutes	75
Mathematics	60 minutes	60
Reading	35 minutes	40
Science	35 minutes	40
Writing (optional)	40 minutes	1 essay

Scoring

Test takers receive a score between 1 and 36 for each of the four subject areas. Those scores are averaged together to give a Composite Score, which is the primary score reported as an "ACT score." The most prestigious schools typically admit students with Composite ACT Scores in the low 30's. Other selective schools typically admit students with scores in the high 20's. Traditional colleges more likely admit students with scores in the low 20's, while community colleges and other more open schools typically accept students with scores in the high teens. In 2015, the average composite score among all test takers (including those not applying to college) was 21.

Recent/Future Developments

In 2015, the Writing section underwent several changes. The allotted time extended 10 minutes (from 30 to 40 minutes) and the scoring changed to a scale from 1 to 36 (as with the other subject and Composite scores), rather than the previous scale from 2 to 12. The test also began asking test takers to give an opinion on a subject in light of three different perspectives provided by the test prompt, Lastly, the ACT began reporting four new "subscores," providing different ways to combine and evaluate the results of the various sections.

Beginning in September 2016, the scoring of the writing section changed back to a 2 to 12 scale.

Biology

Biology is the study of living organisms and the processes that are vital for life. Scientists who study biology study these organisms on a cellular level, individually or as populations, and look at the effects they have on their surrounding environment.

Water

Most cells are primarily composed of water and live in water-rich environments. Since water is such a familiar substance, it is easy to overlook its unique properties. Chemically, water is made up of two hydrogen atoms bonded to one oxygen atom by covalent bonds. The three atoms join to make a V-shaped molecule. Water is a polar molecule, meaning it has an unevenly distributed overall charge due to an unequal sharing of electrons. Due to oxygen's electronegativity and its more substantial positively charged nucleus, hydrogen's electrons are pulled closer to the oxygen. This causes the hydrogen atoms to have a slight positive charge and the oxygen atom to have a slight negative charge. In a glass of water, the molecules constantly interact and link for a fraction of a second due to intermolecular bonding between the slightly positive hydrogen atoms of one molecule and the slightly negative oxygen of a different molecule. These weak intermolecular bonds are called **hydrogen bonds**.

Water has several important qualities, including: cohesive and adhesive behaviors, temperature moderation ability, expansion upon freezing, and diverse use as a solvent.

Cohesion is the interaction of many of the same molecules. In water, cohesion occurs when there is hydrogen bonding between water molecules. Water molecules use this bonding ability to attach to each other and can work against gravity to transport dissolved nutrients to the top of a plant. A network of water-conducting cells can push water from the roots of a plant up to the leaves. Adhesion is the linking of two different substances. Water molecules can form a weak hydrogen bond with, or adhere to, plant cell walls to help fight gravity. The cohesive behavior of water also causes surface tension. If a glass of water is slightly overfull, water can still stand above the rim. This is because of the unique bonding of water molecules at the surface—they bond to each other and to the molecules below them, making it seem like it is covered with an impenetrable film. A raft spider could actually walk across a small body of water due to this surface tension.

Another important property of water is its ability to moderate temperature. Water can moderate the temperature of air by absorbing or releasing stored heat into the air. Water has the distinctive capability of being able to absorb or release large quantities of stored heat while undergoing only a small change in temperature. This is because of the relatively high **specific heat** of water, where specific heat is the amount of heat it takes for one gram of a material to change its temperature by 1 degree Celsius. The specific heat of water is one calorie per gram per degree Celsius, meaning that for each gram of water, it takes one calorie of heat to raise or lower the temperature of water by 1 degree Celsius.

When the temperature of water is reduced to freezing levels, water displays another interesting property: It expands instead of contracts. Most liquids become denser as they freeze because the molecules move around slower and stay closer together. Water molecules, however, form hydrogen bonds with each other as they move together. As the temperature lowers and they begin to move slower, these bonds become harder to break apart. When water freezes into ice, molecules are frozen with hydrogen bonds between them and they take up about 10 percent more volume than in their liquid state. The fact that ice is less dense than water is what makes ice float to the top of a glass of water.

Lastly, the **polarity** of water molecules makes it a versatile solvent. **Ionic compounds**, such as salt, are made up of positively- and negatively-charged atoms, called **cations** and **anions**, respectively. Cations and anions are easily dissolved in water because of their individual attractions to the slight positive charge of the hydrogen atoms or the slight negative charge of the oxygen atoms in water molecules. Water molecules separate the individually charged atoms and shield them from each other so they don't bond to each other again, creating a homogenous solution of the cations and anions. Nonionic compounds, such as sugar, have polar regions, so are easily dissolved in water. For these compounds, the water molecules form hydrogen bonds with the polar regions (hydroxyl groups) to create a homogenous solution. Any substance that is attracted to water is termed **hydrophilic**. Substances that repel water are termed **hydrophobic**.

Biological Molecules

Basic units of organic compounds are often called **monomers**. Repeating units of linked monomers are called **polymers**. The most important large molecules, or polymers, found in all living things can be divided into four categories: carbohydrates, lipids, proteins, and nucleic acids. This may be surprising since there is so much diversity in the outward appearance and physical abilities of living things present on Earth. Carbon (C), hydrogen (H), oxygen (O), nitrogen (N), sulfur (S), and phosphorus (P) are the major elements of most biological molecules. Carbon is a common backbone of large molecules because of its ability to form four covalent bonds.

Carbohydrates

Carbohydrates consist of sugars and polymers of sugars. The simplest sugar are **monosaccharide**, which have the empirical formula of CH_2O. The formula for the monosaccharide glucose, for example, is $C_6H_{12}O_6$. Glucose is an important molecule for cellular respiration, the process of cells extracting energy by breaking bonds through a series of reactions. The individual atoms are then used to rebuild new small molecules. **Polysaccharides** are made up of a few hundred to a few thousand monosaccharides linked together. These larger molecules have two major functions. The first is that they can be stored as starches, such as **glycogen**, and then broken down later for energy. Secondly, they may be used to form strong materials, such as **cellulose**, which is the firm wall that encloses plant cells, and **chitin**, the carbohydrate insects use to build exoskeletons.

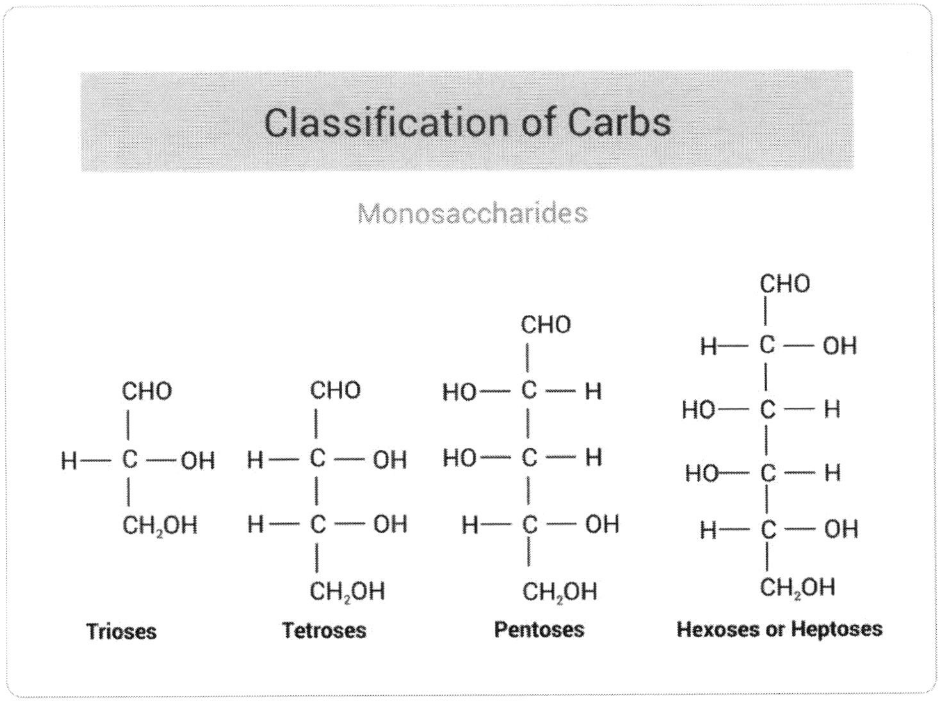

Classification of Carbs

Monosaccharides

Trioses	Tetroses	Pentoses	Hexoses or Heptoses
CHO	CHO	CHO	CHO
H—C—OH	H—C—OH	HO—C—H	H—C—OH
CH_2OH	H—C—OH	HO—C—H	HO—C—H
	CH_2OH	H—C—OH	HO—C—H
		CH_2OH	H—C—OH
			CH_2OH

Lipids

Lipids are a class of biological molecules that are **hydrophobic**, meaning they don't mix well with water. They are mostly made up of large chains of carbon and hydrogen atoms, termed **hydrocarbon chains**. When lipids mix with water, the water molecules bond to each other and exclude the lipids because they are unable to form bonds with the long hydrocarbon chains. The three most important types of lipids are fats, phospholipids, and steroids.

Fats are made up of two types of smaller molecules: glycerol and fatty acids. **Glycerol** is a chain of three carbon atoms, with a **hydroxyl group** attached to each carbon atom. A hydroxyl group is made up of an oxygen and hydrogen atom bonded together. **Fatty acids** are long hydrocarbon chains that have a backbone of sixteen or eighteen carbon atoms. The carbon atom on one end of the fatty acid is part of a **carboxyl group.** A carboxyl group is a carbon atom that uses two of its four bonds to bond to one oxygen atom (double bond) and uses another one of its bonds to link to a hydroxyl group.

Fats are made by joining three fatty acid molecules and one glycerol molecule.

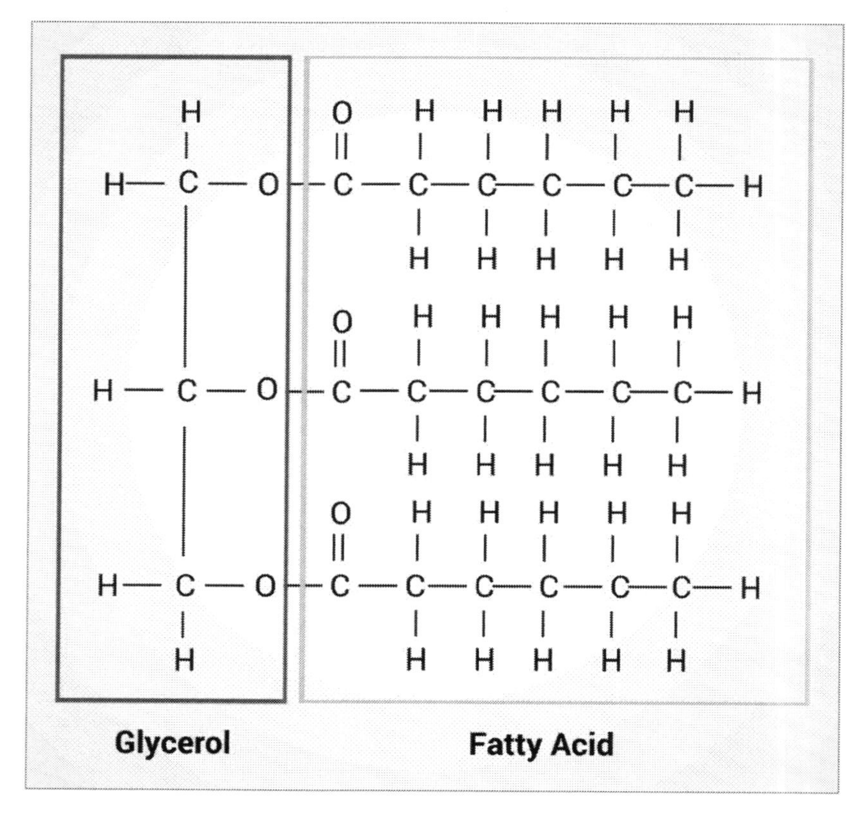

Phospholipids are made of two fatty acid molecules linked to one glycerol molecule. A **phosphate group** is attached to a third hydroxyl group of the glycerol molecule. A phosphate group consists of a phosphate atom connected to four oxygen atoms and has an overall negative charge.

Phospholipids have an interesting structure because their fatty acid tails are hydrophobic, but their phosphate group heads are hydrophilic. When phospholipids mix with water, they create double-layered structures, called **bilayers,** that shield their hydrophobic regions from water molecules. Cell membranes are made of phospholipid bilayers, which allow the cells to mix with aqueous solutions outside and inside, while forming a protective barrier and a semi-permeable membrane around the cell.

Steroids are lipids that consist of four fused carbon rings. The different chemical groups that attach to these rings are what make up the many types of steroids. **Cholesterol** is a common type of steroid found in animal cell membranes. Steroids are mixed in between the phospholipid bilayer and help maintain the structure of the membrane and aids in cell signaling.

Proteins

Proteins are essential for most all functions in living beings. The term *protein* is derived from the Greek word *proteios*, meaning *first* or *primary*. All proteins are made from a set of twenty amino acids that are linked in unbranched polymers. The combinations are numerous, which accounts for the diversity of proteins. Amino acids are linked by peptide bonds, while polymers of amino acids are called **polypeptides**. These polypeptides, either individually or in linked combination with each other, fold up to form coils of biologically-functional molecules, called proteins.

There are four levels of protein structure: primary, secondary, tertiary, and quaternary. The **primary structure** is the sequence of amino acids, similar to the letters in a long word. The **secondary structure** is beta sheets, or alpha helices, formed by hydrogen bonding between the polar regions of the polypeptide backbone. **Tertiary structure** is the overall shape of the molecule that results from the interactions between the side chains linked to the polypeptide backbone. **Quaternary structure** is the overall protein structure that occurs when a protein is made up of two or more polypeptide chains.

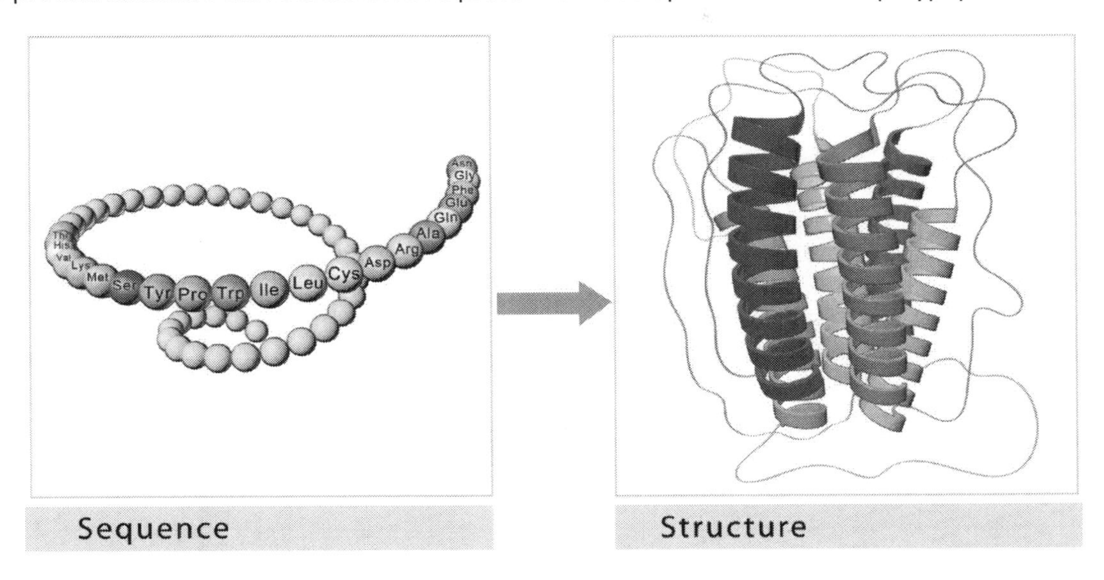

Nucleic Acids

Nucleic acids can also be called **polynucleotides** because they are made up of chains of monomers called **nucleotides.** Nucleotides consist of a five-carbon sugar, a nitrogen-containing base, and a phosphate group. There are two types of nucleic acids: **deoxyribonucleic acid (DNA)** and **ribonucleic acid (RNA)**. Both DNA and RNA enable living organisms to pass on their genetic information and complex components to subsequent generations. While DNA is made up of two strands of nucleotides coiled together in a double-helix structure, RNA is made up of a single strand of nucleotides that folds onto itself.

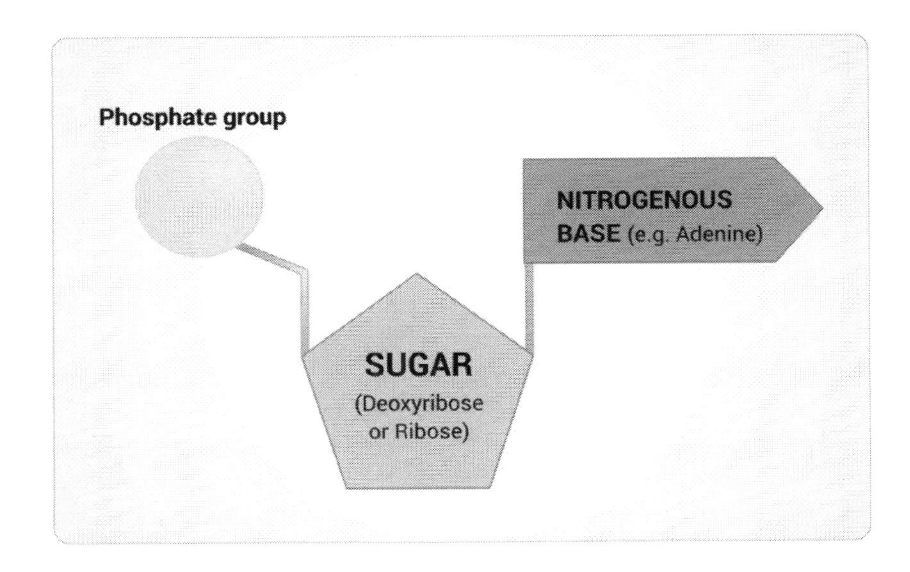

Metabolism

Metabolism is the set of chemical processes that occur within a cell for the maintenance of life. It includes both the synthesizing and breaking down of substances. A metabolic pathway begins with a molecule and ends with a specific product after going through a series of reactions, often involving an enzyme at each step. An **enzyme** is a protein that aids in the reaction. **Catabolic pathways** are metabolic pathways in which energy is released by complex molecules being broken down into simpler molecules. Contrast to catabolic pathways are **anabolic pathways**, which use energy to build complex molecules out of simple molecules. With cell metabolism, remember the **first law of thermodynamics**: Energy can be transformed, but it cannot be created or destroyed. Therefore, the energy released in a cell by a catabolic pathway is used up in anabolic pathways.

The reactions that occur within metabolic pathways are classified as either exergonic reactions or endergonic reactions. **Exergonic reactions** end in a release of free energy, while **endergonic reactions** absorb free energy from its surroundings. **Free energy** is the portion of energy in a system, such as a living cell, that can be used to perform work, such as a chemical reaction. It is denoted as the capital letter G and the change in free energy from a reaction or set of reactions is denoted as delta G (ΔG). When reactions do not require an input of energy, they are said to occur spontaneously. Exergonic reactions are considered spontaneous because they result in a negative delta G ($-\Delta G$), where the products of the reaction have less free energy within them than the reactants. Endergonic reactions require an input of energy and result in a positive delta G ($+\Delta G$), with the products of the reaction containing more free energy than the individual reactants. When a system no longer has free energy to do work, it has reached **equilibrium**. Since cells always work, they are no longer alive if they reach equilibrium.

Cells balance their energy resources by using the energy from exergonic reactions to drive endergonic reactions forward, a process called **energy coupling**. **Adenosine triphosphate**, or ATP, is a molecule that is an immediate source of energy for cellular work. When it is broken down, it releases energy used in endergonic reactions and anabolic pathways. ATP breaks down into adenosine diphosphate, or ADP, and a separate phosphate group, releasing energy in an exergonic reaction. As ATP is used up by reactions, it is also regenerated by having a new phosphate group added onto the ADP products within the cell in an endergonic reaction.

Enzymes are special proteins that help speed up metabolic reactions and pathways. They do not change the overall free energy release or consumption of reactions; they just make the reactions occur more quickly as it lowers the activation energy required. Enzymes are designed to act only on specific substrates. Their physical shape fits snugly onto their matched substrates, so enzymes only speed up reactions that contain the substrates to which they are matched.

The Cell

Cells are the basic structural and functional unit of all organisms. They are the smallest unit of matter that is living. While there are many single-celled organisms, most biological organisms are more complex and made up of many different types of cells. There are two distinct types of cells: prokaryotic and eukaryotic. **Prokaryotic cells** include bacteria, while **eukaryotic cells** include animal and plant cells. Both types of cells are enclosed by a cell membrane, which is selectively permeable. Selective permeability means essentially that it is a gatekeeper, allowing certain molecules and ions in and out, and keeping unwanted ones at bay, at least until they are ready for use. Both contain ribosomes, which are complexes that make protein inside the cell, and DNA. One major difference is that the DNA in eukaryotic cells are enclosed in a membrane-bound **nucleus**, where in prokaryotic cells, DNA is in the **nucleoid**, a region that is not enclosed by a membrane. Another major difference is that eukaryotic cells contain **organelles,** which are membrane-enclosed structures, each with a specific function, while prokaryotic cells do not have organelles.

Organelles Found in Eukaryotic Cells

The following cell organelles are found in both animal and plant cells unless otherwise noted:

Nucleus: The nucleus consists of three parts: nuclear envelope, nucleolus, and chromatin. The **nuclear envelope** is the double membrane that surrounds the nucleus and separates its contents from the rest of the cell. It is porous so substances can pass back and forth between the nucleus and the other parts of the cell. It is also continuous, with the endoplasmic reticulum that is present within the cytosol of the cell. The **nucleolus** is in charge of producing ribosomes. **Chromosomes are comprised of tightly coiled proteins, RNA, and DNA and are collectively called chromatin.**

Endoplasmic Reticulum (ER): The ER is a network of membranous sacs and tubes responsible for membrane synthesis and other metabolic and synthetic activities of the cell. There are two types of ER, rough and smooth. Rough ER is lined with ribosomes and is the location of protein synthesis. This provides a separate compartment for site-specific protein synthesis and is important for the intracellular transport of proteins. Smooth ER does not contain ribosomes and is the location of lipid synthesis.

Flagellum: The flagellum is found in protists and animal cells. It is a cluster of microtubules projected out of the plasma membrane and aids in cell motility.

Centrosome: The centrosome is the area of the cell where microtubules are created and organized for mitosis. Each centrosome contains two **centrioles.**

Cytoskeleton: The cytoskeleton in animal cells is made up of microfilaments, intermediate filaments, and microtubules. In plant cells, the cytoskeleton is made up of only microfilaments and microtubules. These structures reinforce the cell's shape and aid in cell movements.

Microvilli: Microvilli are found only in animal cells. They are protrusions in the cell membrane that increase the cell's surface area. They have a variety of functions, including absorption, secretion, and cellular adhesion.

Peroxisome: A peroxisome contains enzymes that are involved in many of the cell's metabolic functions, one of the most important being the breakdown of fatty acid chains. It produces hydrogen peroxide as a by-product of these processes and then converts the hydrogen peroxide to water.

Mitochondrion: The mitochondrion, considered the cell's powerhouse, is one of the most important structures for maintaining regular cell function. It is where cellular respiration occurs and where most of the cell's ATP is generated.

Lysosome: Lysosomes are found exclusively in animal cells. They are responsible for digestion and can hydrolyze macromolecules.

Golgi Apparatus: The Golgi apparatus is responsible for synthesizing, modifying, sorting, transporting, and secreting cell products. Because of its large size, it was one of the first organelles studied in detail.

Ribosomes: Ribosomes are found either free in the cytosol, bound to the rough ER, or bound to the nuclear envelope. They are also found in prokaryotes. Ribosomes make up a complex that forms proteins within the cell.

Plasmodesmata: Found only in plant cells, plasmodesmata are cytoplasmic channels, or tunnels, that go through the cell wall and connect the cytoplasm of adjacent cells.

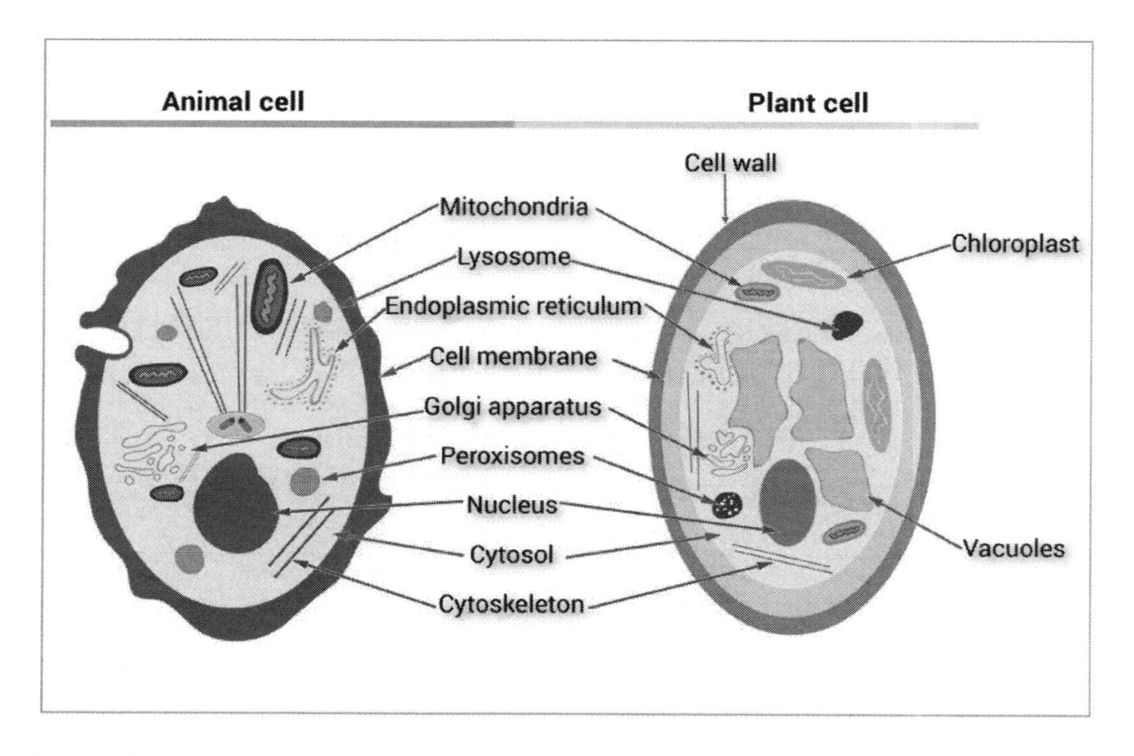

Chloroplast: Chloroplasts are found in protists, such as algae and plant cells. It is responsible for photosynthesis, which is the process of converting sunlight to chemical energy that is stored and used later to drive cellular activities.

Central Vacuole: A central vacuole is found only in plant cells, and is responsible for storage, breakdown of waste products, and hydrolysis of macromolecules.

Plasma Membrane: The plasma membrane is a phospholipid bilayer that encloses the cell. It is also found in prokaryotes.

Cell Wall: Cell walls are present in fungi, plant cells, and some protists. The cell wall is made up of strong fibrous substances, including cellulose (plants), chitin (fungi) and other polysaccharides, and protein. It is a layer outside of the plasma membrane that protects the cell from mechanical damage and helps maintain the cell's shape.

Cellular Respiration

Cellular respiration is a set of metabolic processes that converts energy from nutrients into ATP. Respiration can either occur aerobically, using oxygen, or anaerobically, without oxygen. While prokaryotic cells carry out respiration in the cytosol, most of the respiration in eukaryotic cells occurs in the mitochondria.

Aerobic Respiration

There are three main steps in aerobic cellular respiration: glycolysis, the citric acid cycle (also known as the Krebs cycle), and oxidative phosphorylation. **Glycolysis** is an essential metabolic pathway that converts glucose to pyruvate and allows for cellular respiration to occur. It does not require oxygen to be present. Glucose is a common molecule used for energy production in cells. During glycolysis, two three-carbon sugars are generated from the splitting of a glucose molecule. These smaller sugars are then converted into pyruvate molecules via oxidation and atom rearrangement. Glycolysis requires two ATP molecules to drive the process forward, but the end product of the process has four ATP molecules, for a net production of two ATP molecules. Also, two reduced nicotinamide adenine dinucleotide (NADH) molecules are created from when the electron carrier oxidized nicotinamide adenine dinucleotide (NAD+) peels off two electrons and a hydrogen atom.

In aerobically-respiring eukaryotic cells, the pyruvate molecules then enter the mitochondrion. Pyruvate is oxidized and converted into a compound called acetyl-CoA. This molecule enters the **citric acid cycle** to begin the process of aerobic respiration.

The citric acid cycle has eight steps. Remember that glycolysis produces two pyruvate molecules from each glucose molecule. Each pyruvate molecule oxidizes into a single acetyl-CoA molecule, which then enters the citric acid cycle. Therefore, two citric acid cycles can be completed and twice the number of ATP molecules are generated per glucose molecule.

Eight Steps of the Citric Acid Cycle

Step 1: Acetyl-CoA adds a two-carbon acetyl group to an oxaloacetate molecule and produces one citrate molecule.

Step 2: Citrate is converted to its isomer isocitrate by removing one water molecule and adding a new water molecule in a different configuration.

Step 3: Isocitrate is oxidized and converted to α-ketoglutarate. A carbon dioxide (CO_2) molecule is released and one NAD+ molecule is converted to NADH.

Step 4: α-Ketoglutarate is converted to succinyl-CoA. Another carbon dioxide molecule is released and another NAD+ molecule is converted to NADH.

Step 5: Succinyl-CoA becomes succinate by the addition of a phosphate group to the cycle. The oxygen molecule of the phosphate group attaches to the succinyl-CoA molecule and the CoA group is released. The rest of the phosphate group transfers to a guanosine diphosphate (GDP) molecule, converting it to guanosine triphosphate (GTP). GTP acts similarly to ATP and can actually be used to generate an ATP molecule at this step.

Step 6: Succinate is converted to fumarate by losing two hydrogen atoms. The hydrogen atoms join a flavin adenine dinucleotide (FAD) molecule, converting it to $FADH_2$, which is a hydroquinone form.

Step 7: A water molecule is added to the cycle and converts fumarate to malate.

Step 8: Malate is oxidized and converted to oxaloacetate. One lost hydrogen atom is added to an NAD molecule to create NADH. The oxaloacetate generated here then enters back into step one of the cycle.

At the end of glycolysis and the citric acid cycles, four ATP molecules have been generated. The NADH and $FADH_2$ molecules are used as energy to drive the next step of oxidative phosphorylation.

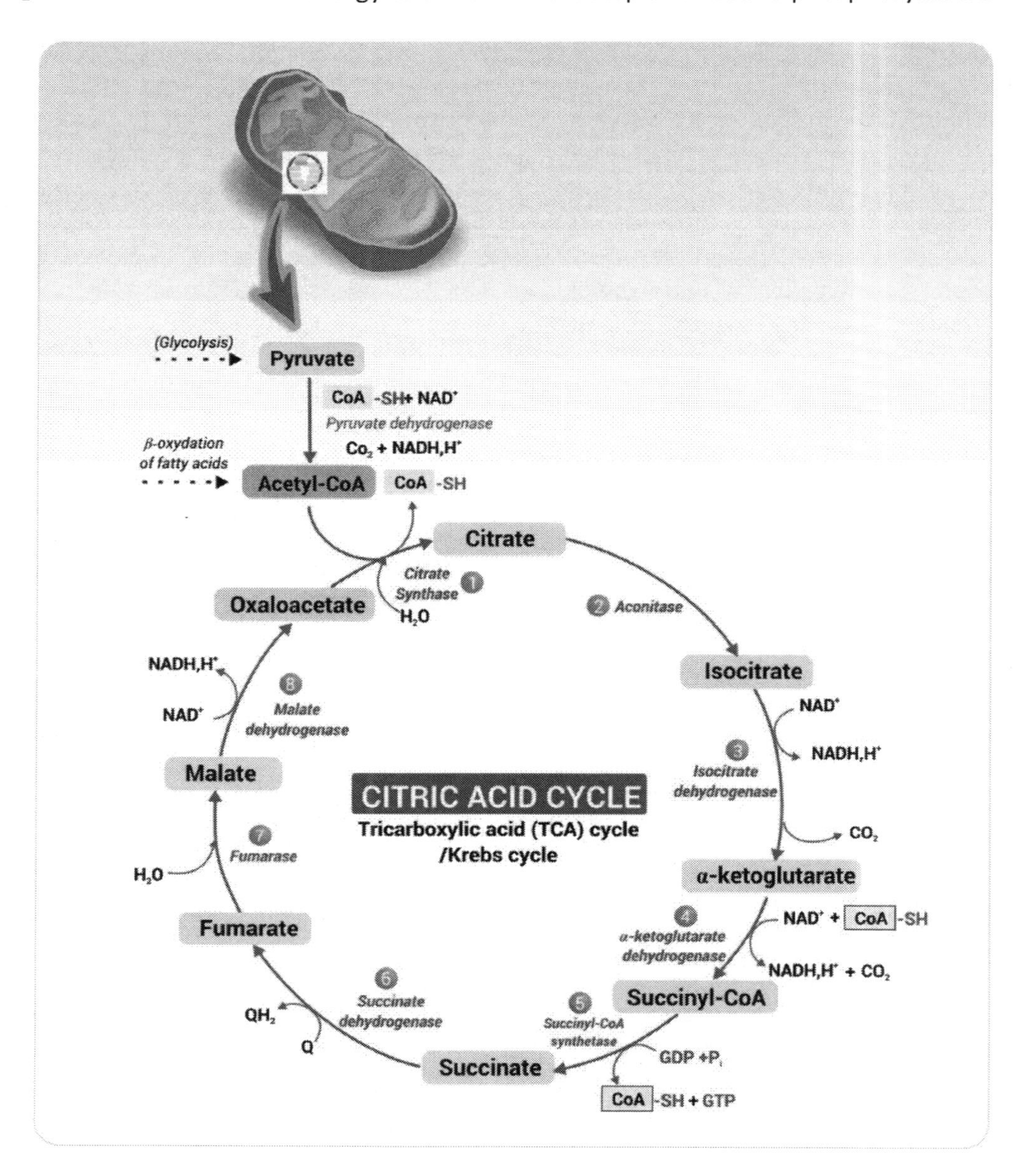

Oxidative Phosphorylation

Oxidative phosphorylation includes two steps: the electron transport chain and chemiosmosis. The inner mitochondrial membrane has four protein complexes, sequenced I to IV, used to transport protons and electrons through the inner mitochondrial matrix. Two electrons and a proton (H+) are passed from each NADH and $FADH_2$ to these channel proteins, pumping the hydrogen ions to the inner-membrane space using energy from the high-energy electrons to create a concentration gradient. NADH and $FADH_2$ also drop their high-energy electrons to the electron transport chain. NAD+ and FAD molecules in the mitochondrial matrix return to the Krebs cycle to pick up materials for the next delivery.

From here, two processes happen simultaneously:

1. **Electron Transport Chain:** In addition to complexes I to IV, there are two mobile electron carriers present in the inner mitochondrial membrane, called **ubiquinone** and **cytochrome C.** At the end of this transport chain, electrons are accepted by an O_2 molecule in the matrix, and water is formed with the addition of two hydrogen atoms from chemiosmosis.

2. **Chemiosmosis:** This occurs in an ATP synthase complex that sits next to the four electron transporting complexes. ATP synthase uses **facilitated diffusion** (passive transport) to deliver protons across the concentration gradient from the inner mitochondrial membrane to the matrix. As the protons travel, the ATP synthase protein physically spins, and the kinetic energy generated is invested into phosphorylation of ADP molecules to generate ATP. Oxidative phosphorylation produces twenty-six to twenty-eight ATP molecules, bringing the total number of ATP generated through glycolysis and cellular respiration to thirty to thirty-two molecules.

Anaerobic Respiration

Some organisms do not live in oxygen-rich environments and must find alternate methods of respiration. Anaerobic respiration occurs in certain prokaryotic organisms. They utilize an electron transport chain similar to the aerobic respiration pathway; however, the terminal acceptor molecule is an electronegative substance that is not O_2. Some bacteria, for example, use the sulfate ion (SO_4^{2-}) as the final electron accepting molecule and the resulting byproduct is hydrogen sulfide (H_2S) instead of water.

Muscle cells that reach anaerobic threshold go through lactic acid respiration, while yeasts go through alcohol fermentation. Both processes only make two ATP.

Photosynthesis

Photosynthesis is the process of converting light energy into chemical energy that is then stored in sugar and other organic molecules. It can be divided into two stages: the light-dependent reactions and the Calvin cycle. In plants, the photosynthetic process takes place in the chloroplast. Inside the chloroplast are membranous sacs, called **thylakoids**. Chlorophyll is a green pigment that lives in the thylakoid membranes and absorbs the light energy, starting the process of photosynthesis. The **Calvin cycle** takes place in the **stroma,** or inner space, of the chloroplasts. The complex series of reactions that take place in photosynthesis can be simplified into the following equation: $6CO_2 + 12H_2O$ + Light Energy $\rightarrow C_6H_{12}O_6 + 6O_2 + 6H_2O$. Basically, carbon dioxide and water mix with light energy inside the chloroplast to produce organic molecules, oxygen, and water. Note that water is on both sides of the equation. Twelve water molecules are consumed during this process and six water molecules are newly formed as byproducts.

The Light Reactions

During the **light reactions**, chlorophyll molecules absorb light energy, or solar energy. In the thylakoid membrane, chlorophyll molecules, together with other small molecules and proteins, form photosystems, which are made up of a reaction-center complex surrounded by a light-harvesting complex. In the first step of photosynthesis, the light-harvesting complex from photosystem II (PSII) absorbs a photon from light, passes the photon from one pigment molecule to another within itself, and then transfers it to the reaction-center complex. Inside the reaction-center complex, the energy from the photon enables a special pair of chlorophyll *a* molecules to release two electrons. These two electrons are then accepted by a primary electron acceptor molecule. Simultaneously, a water molecule is split into two hydrogen atoms, two electrons and one oxygen atom. The two electrons are transferred one by one to the chlorophyll *a* molecules, replacing their released electrons. The released electrons are then transported down an electron transport chain by attaching to the electron carrier plastoquinone (Pq), a cytochrome complex, and then a protein called plastocyanin (Pc) before they reach photosystem I (PS I). As the electrons pass through the cytochrome complex, protons are pumped into the thylakoid space, providing the concentration gradient that will eventually travel through ATP synthase to make ATP (like in aerobic respiration). PS I absorbs photons from light, similar to PS II. However, the electrons that are released from the chlorophyll *a* molecules in PS I are replaced by the electrons coming down the electron transport chain (from PS II). A primary electron acceptor molecule accepts the released electrons in PS I and passes the electrons onto another electron transport chain involving the protein ferredoxin (Fd). In the final steps of the light reactions, electrons are transferred from Fd to Nicotinamide adenine dinucleotide phosphate (NADP+) with the help of the enzyme NADP+ reductase and NADPH is produced. The ATP and nicotinamide adenine dinucleotide phosphate-oxidase (NADPH) produced from the light reactions are used as energy to form organic molecules in the Calvin cycle.

The Calvin Cycle

There are three phases in the Calvin cycle: carbon fixation, reduction, and regeneration of the CO_2 acceptor. **Carbon fixation** is when the first carbon molecule is introduced into the cycle, when CO_2 from the air is absorbed by the chloroplast. Each CO_2 molecule enters the cycle and attaches to ribulose bisphosphate (RuBP), a five-carbon sugar. The enzyme RuBP carboxylase-oxygenase, also known as rubisco, catalyzes this reaction. Next, two three-carbon 3-phosphoglycerate sugar molecules are formed immediately from the splitting of the six-carbon sugar.

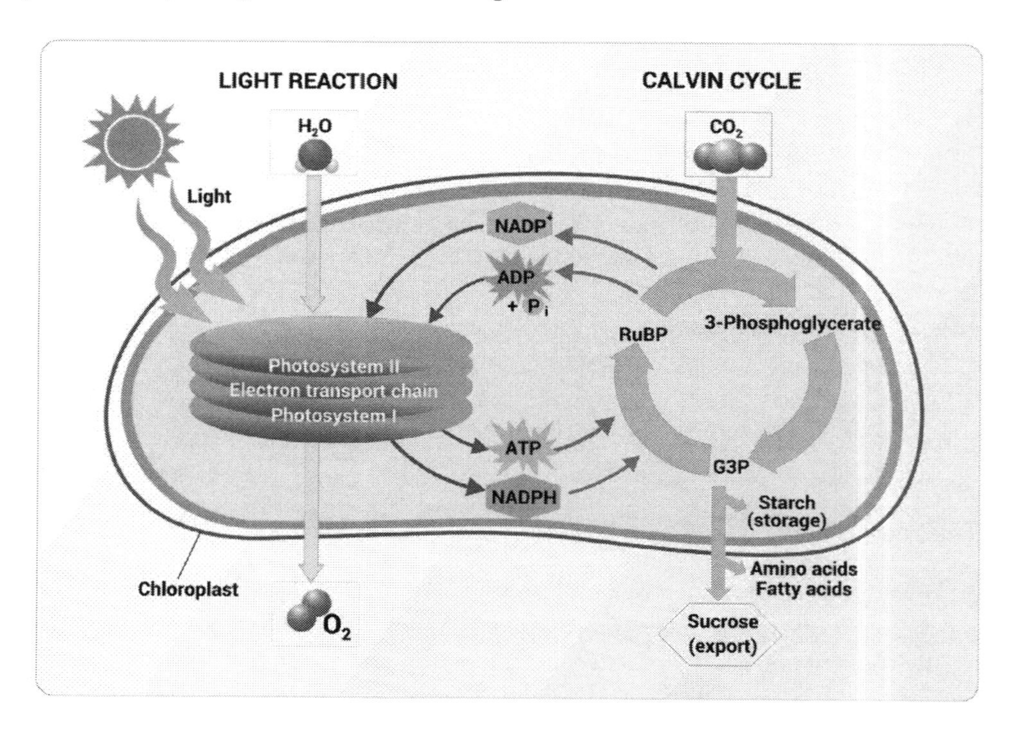

Next, during the **reduction** phase, an ATP molecule is reduced to ADP and the phosphate group attaches to 3-phosphoglycerate, forming 1,3-bisphosphoglycerate. An NADPH molecule then donates two high-energy electrons to the newly formed 1,3-bisphosphate, causing it to lose the phosphate group and become glyceraldehyde 3-phosphate (G3P), which is a high-energy sugar molecule. At this point in the cycle, one G3P molecule exits the cycle and is used by the plant. However, to regenerate RuBP molecules, which are the CO_2 acceptors in the cycle, five G3P molecules continue in the cycle. It takes three turns of the cycle and three CO_2 molecules entering the cycle to form one G3P molecule.

In the final phase of the Calvin cycle, three RuBP molecules are formed from the rearrangement of the carbon skeletons of five G3P molecules. It is a complex process that involves the reduction of three ATP molecules. At the end of the process, RuBP molecules are again ready to enter the first phase and accept CO_2 molecules.

Although the Calvin cycle is not dependent on light energy, both steps of photosynthesis usually occur during daylight, as the Calvin cycle is dependent upon the ATP and NADPH produced by the light reactions, because that energy can be invested into bonds to create high-energy sugars. The Calvin cycle invests nine ATP molecules and six NADPH molecules into every one molecule of G3P that it produces. The G3P that is produced can be used as the starting material to build larger organic compounds, such as glucose.

Cellular Reproduction

Cellular reproduction is the process that cells use to divide into two new cells. The ability of a multi-cellular organism to generate new cells to replace dying and damaged cells is vital for sustaining its life. There are two processes by which a cell can divide: mitosis and meiosis. In **mitosis,** the daughter cells produced from parental cell division are identical to each other and the parent. **Meiosis** produces genetically unique haploid cells due to two stages of cell division. Meiosis produces **haploid** cells, or **gametes** (sperm and egg cells), which only have one set of chromosomes. Humans are **diploid** because we have two sets of chromosomes – one from each parent. **Somatic** (body) cells are all diploid and are produced via mitosis.

Mitosis

Mitosis is the division of the genetic material in the nucleus of a cell, and is immediately followed by **cytokinesis**, which is the division of the cytoplasm of the cell. The two processes make up the mitotic phase of the cell cycle. Mitosis can be broken down into five stages: prophase, prometaphase, metaphase, anaphase, and telophase. Mitosis is preceded by **interphase**, where the cell spends the majority of its life while growing and replicating its DNA.

Prophase: During this phase, the mitotic spindles begin to form. They are made up of centrosomes and microtubules. As the microtubules lengthen, the centrosomes move farther away from each other. The nucleolus disappears and the chromatin fibers begin to coil up and form chromosomes. Two sister **chromatids**, which are two identical copies of one chromosome, are joined together at the centromere.

Prometaphase: The nuclear envelope begins to break down and the microtubules enter the nuclear area. Each pair of chromatin fibers develops a **kinetochore**, which is a specialized protein structure in the middle of the adjoined fibers. The chromosomes are further condensed.

Metaphase: The microtubules are stretched across the cell and the centrosomes are at opposite ends of the cell. The chromosomes align at the metaphase plate, which is a plane that is exactly between the two centrosomes. The centromere of each chromosome is attached to the kinetochore microtubules that are stretching from each centrosome to the metaphase plate.

Anaphase: The sister chromatids break apart, forming individual chromosomes. The two daughter chromosomes move to opposite ends of the cell. The microtubules shorten toward opposite ends of the cell as well. The cell elongates and, by the end of this phase, there is a complete set of chromosomes at each end of the cell.

Telophase: Two nuclei form at each end of the cell and nuclear envelopes begin to form around each nucleus. The nucleoli reappear and the chromosomes become less condensed. The microtubules are broken down by the cell and mitosis is complete.

Cytokinesis divides the cytoplasm by pinching off the cytoplasm, forming a cleavage furrow, and the two daughter cells then enter interphase, completing the cycle.

Plant cell mitosis is similar except that it lacks centromeres, and instead has a microtubule organizing center. Cytokinesis occurs with the formation of a cell plate.

Meiosis

Meiosis is a type of cell division in which the parent cell has twice as many sets of chromosomes as the daughter cells into which it divides. Although the first stage of meiosis involves the duplication of chromosomes, similar to that of mitosis, the parent cell in meiosis divides into four cells, as opposed to the two produced in mitosis.

Meiosis has the same phases as mitosis, except that they occur twice: once in meiosis I and again in meiosis II. The diploid parent has two sets of homologous chromosomes, one set from each parent. During meiosis I, each chromosome set goes through a process called **crossing over**, which jumbles up the genes on each chromatid. In anaphase one, the separated chromosomes are no longer identical and, once the chromosomes pull apart, each daughter cell is haploid (one set of chromosomes with two non-identical sister chromatids). Next, during meiosis II, the two intermediate daughter cells divide again, separating the chromatids, producing a total of four total haploid cells that each contains one set of chromosomes.

Genetics

Genetics is the study of heredity, which is the transmission of traits from one generation to the next, and hereditary variation. The chromosomes passed from parent to child contain hereditary information in the form of genes. Each gene has specific sequences of DNA that encode proteins, start pathways, and result in inherited traits. In the human life cycle, one haploid sperm cell joins one haploid egg cell to form a diploid cell. The diploid cell is the zygote, the first cell of the new organism, and from then on mitosis takes over and nine months later, there is a fully developed human that has billions of identical cells.

The monk Gregor Mendel is referred to as the father of genetics. In the 1860s, Mendel came up with one of the first models of inheritance, using peapods with different traits in the garden at his abbey to test his theory and develop his model. His model included three laws to determine which traits are inherited; his theories still apply today, after genetics has been studied more in depth.

1. The **Law of Dominance:** Each characteristic has two versions that can be inherited. The gene that encodes for the characteristic has two variations, or alleles, and one is dominant over the other.

2. The **Law of Segregation:** When two parent cells form daughter cells, the alleles segregate and each daughter cell only inherits one of the alleles from each parent.

3. The **Law of Independent Assortment:** Different traits are inherited independent of one another because in metaphase, the set of chromosomes line up in random fashion – mom's set of chromosomes do not line up all on the left or right, there is a random mix.

Dominant and Recessive Traits

Each gene has two **alleles**, one inherited from each parent. **Dominant alleles** are noted in capital letters (A) and **recessive alleles** are noted in lower case letters (a). There are three possible combinations of alleles among dominant and recessive alleles: AA, Aa (known as a heterozygote), and aa. Dominant alleles, when mixed with recessive alleles, will mask the recessive trait. The recessive trait would only appear as the phenotype when the allele combination is aa because a dominant allele is not present to mask it.

Although most genes follow the standard dominant/recessive rules, there are some genes that defy them. Examples include cases of co-dominance, multiple alleles, incomplete dominance, sex-linked traits, and polygenic inheritance.

In cases of **co-dominance**, both alleles are expressed equally. For example, blood type has three alleles: I^A, I^B, and i. I^A and I^B are both dominant to i, but co-dominant with each other. An I^AI^B has AB blood. With incomplete dominance, the allele combination Aa actually makes a third phenotype. An example: certain flowers can be red (AA), white (aa), or pink (Aa).

Punnett Square

For simple genetic combinations, a **Punnett square** can be used to assess the phenotypes of subsequent generations. In a 2 x 2 cell square, one parent's alleles are set up in columns and the other parent's alleles are in rows. The resulting allele combinations are shown in the four internal cells.

Mutations

Genetic **mutations** occur when there is a permanent alteration in the DNA sequence that codes for a specific gene. They can be small, affecting only one base pair, or large, affecting many genes on a chromosome. Mutations are classified as either hereditary, which means they were also present in the parent gene, or acquired, meaning they occurred after the genes were passed down from the parents. Although mutations are not common, they are an important aspect of genetics and variation in the general population.

DNA

DNA is made of nucleotide and contains the genetic information of a living organism. It consists of two polynucleotide strands that are twisted and linked together in a double-helix structure. The polynucleotide strands are made up of four nitrogenous bases: adenine (A), thymine (T), guanine (G), and cytosine (C). Adenine and guanine are purines while thymine and cytosine are pyrimidines. These bases have specific pairings of A with T, and G with C. The bases are ordered so that these specific pairings will occur when the two polynucleotide strands coil together to form a DNA molecule. The two strands of DNA are described as antiparallel because one strand runs 5′ $\longrightarrow$ 3′ while the other strand of the helix runs 3′ $\longrightarrow$ 5′.

Before chromosome replication and cell division can occur, DNA replication must happen in interphase. There are specific base pair sequences on DNA, called origins of replication, where DNA replication begins. The proteins that begin the replication process attach to this site and begin separating the two strands and creating a replication bubble. Each end of the bubble has a replication fork, which is a Y-shaped area of the DNA that is being unwound. Several types of proteins are important to the beginning of DNA replication. **Helicases** are enzymes responsible for untwisting the two strands at the replication fork. Single-strand binding proteins bind to the separated strands so that they do not join back together during the replication process. While part of the DNA is unwound, the remainder of the molecule becomes even more twisted in response. Topoisomerase enzymes help relieve this strain by breaking, untwisting, and rejoining the DNA strands.

Once the DNA strand is unwound, an initial primer chain of RNA from the enzyme primase is made to start replication. Replication of DNA can only occur in the 5' → 3' direction. Therefore, during replication, one strand of the DNA template creates the leading strand in the 5' → 3' direction and the other strand creates the lagging strand. While the leading strand is created efficiently and in one piece, the lagging strand is generated in fragments, called **Okazaki fragments**, then are pieced together to form a complete strand by DNA ligase. Following the primer chain of RNA, DNA polymerases are the enzymes responsible for extending the DNA chains by adding on base pairs.

Chemistry

Scientific Notation, the Metric System, and Temperature Scales

Scientific Notation

Scientific notation is the conversion of extremely small or large numbers into a format that is easier to comprehend and manipulate. It changes the number into a product of two separate numbers: a digit term and an exponential term.

Scientific notation = digit term x exponential term

To put a number into scientific notation, one should use the following steps:

- Move the decimal point to after the first non-zero number to find the digit number.
- Count how many places the decimal point was moved in step 1.
- Determine if the exponent is positive or negative.
- Create an exponential term using the information from steps 2 and 3.
- Combine the digit term and exponential term to get scientific notation.

For example, to put 0.0000098 into scientific notation, the decimal should be moved so that it lies between the last two numbers: 000009.8. This creates the digit number:

9.8

Next, the number of places that the decimal point moved is determined; to get between the 9 and the 8, the decimal was moved six places to the right. It may be helpful to remember that a decimal moved to the right creates a negative exponent, and a decimal moved to the left creates a positive exponent. Because the decimal was moved six places to the right, the exponent is negative.

Now, the exponential term can be created by using the base 10 (this is *always* the base in scientific notation) and the number of places moved as the exponent, in this case:

$$10^{-6}$$

Finally, the digit term and the exponential term can be combined as a product. Therefore, the scientific notation for the number 0.0000098 is:

$$9.8 \times 10^{-6}$$

Standard vs. Metric Systems

The measuring system used today in the United States developed from the British units of measurement during colonial times. The most typically used units in this customary system are those used to measure weight, liquid volume, and length, whose common units are found below. In the customary system, the basic unit for measuring weight is the ounce (oz); there are 16 ounces (oz) in 1 pound (lb) and 2000 pounds in 1 ton. The basic unit for measuring liquid volume is the ounce (oz); 1 ounce is equal to 2 tablespoons (tbsp) or 6 teaspoons (tsp), and there are 8 ounces in 1 cup, 2 cups in 1 pint (pt), 2 pints in 1 quart (qt), and 4 quarts in 1 gallon (gal). For measurements of length, the inch (in) is the base unit; 12 inches make up 1 foot (ft), 3 feet make up 1 yard (yd), and 5280 feet make up 1 mile (mi). However, as there are only a set number of units in the customary system, with extremely large or extremely small amounts of material, the numbers can become awkward and difficult to compare.

Common Customary Measurements		
Length	**Weight**	**Capacity**
1 foot = 12 inches	1 pound = 16 ounces	1 cup = 8 fluid ounces
1 yard = 3 feet	1 ton = 2,000 pounds	1 pint = 2 cups
1 yard = 36 inches		1 quart = 2 pints
1 mile = 1,760 yards		1 quart = 4 cups
1 mile = 5,280 feet		1 gallon = 4 quarts
		1 gallon = 16 cups

Aside from the United States, most countries in the world have adopted the **metric system** embodied in the International System of Units (SI). The three main SI base units used in the metric system are the meter (m), the kilogram (kg), and the liter (L); meters measure length, kilograms measure mass, and liters measure volume.

These three units can use different prefixes, which indicate larger or smaller versions of the unit by powers of ten. This can be thought of as making a new unit which is sized by multiplying the original unit in size by a factor.

These prefixes and associated factors are:

Metric Prefixes			
Prefix	**Symbol**	**Multiplier**	**Exponential**
kilo	k	1,000	10^{3}
hecto	h	100	10^{2}
deca	da	10	10^{1}
no prefix		1	10^{0}
deci	d	0.1	10^{-1}
centi	c	0.01	10^{-2}
milli	m	0.001	10^{-3}

The correct prefix is then attached to the base. Some examples:

1 milliliter equals .001 liters.

1 kilogram equals 1,000 grams.

Some units of measure are represented as square or cubic units depending on the solution. For example, perimeter is measured in units, area is measured in square units, and volume is measured in cubic units.

Also, be sure to use the most appropriate unit for the thing being measured. A building's height might be measured in feet or meters while the length of a nail might be measured in inches or centimeters. Additionally, for SI units, the prefix should be chosen to provide the most succinct available value. For example, the mass of a bag of fruit would likely be measured in kilograms rather than grams or milligrams, and the length of a bacteria cell would likely be measured in micrometers rather than centimeters or kilometers.

Temperature Scales

There are three main temperature scales used in science. The scale most often used in the United States is the **Fahrenheit** scale. This scale is based on the measurement of water freezing at 32^{0} F and water boiling at 212^{0} F. The **Celsius** scale uses 0^{0} C as the temperature for water freezing and 100^{0} C for water boiling. The Celsius scale is the most widely used in the scientific community. The accepted measurement by the International System of Units (from the French Système international d'unités), or SI, for temperature is the Kelvin scale. This is the scale employed in thermodynamics, since its zero is the basis for absolute zero, or the unattainable temperature, when matter no longer exhibits degradation.

The conversions between the temperature scales are as follows:

^{0}Fahrenheit to ^{0}Celsius: $^{0}C = \frac{5}{9}(^{0}F - 32)$

^{0}Celsius to ^{0}Fahrenheit: $^{0}F = \frac{9}{5}(^{0}C) + 32$

^{0}Celsius to Kelvin: $K = {}^{0}C + 273.15$

Atomic Structure and the Periodic Table

Atomic Structure

The structure of an atom has two major components: the atomic nucleus and the atomic shells (also known as orbits). The **nucleus** is found in the center of an atom. The three major subatomic particles are protons, neutrons, and electrons and are found in the atomic nucleus and shells.

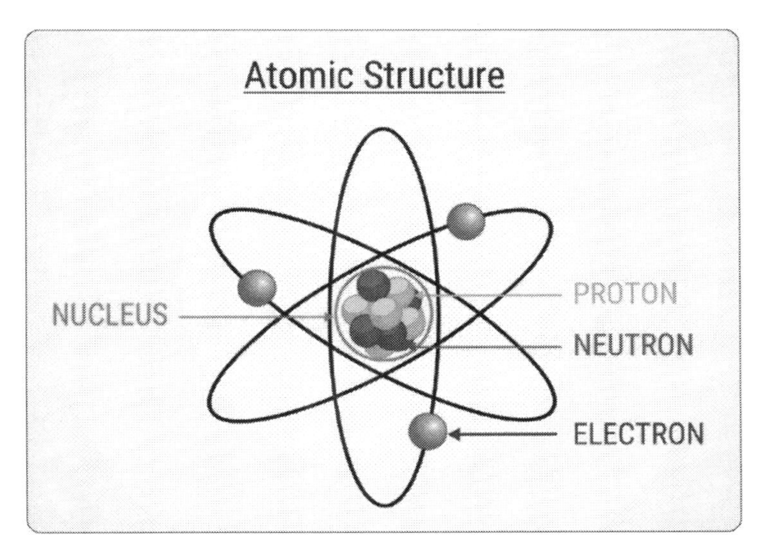

Protons are found in the atomic nucleus and are positively charged particles. The addition or removal of protons from an atom's nucleus creates an entirely different element. **Neutrons** are also found in the atomic nucleus and are neutral particles, meaning they have no net electrical charge. The addition or removal of neutrons from an atom's nucleus does not create a different element but instead creates a lighter or heavier form of that element called an isotope. **Electrons** are found orbiting in the atomic shells around the nucleus and are negatively charged particles. A proton or a neutron has nearly 2,000 times the mass of an electron.

Electrons orbit the nucleus in atomic shells, or electron clouds. For example, the first atomic shell can accommodate two electrons, the second atomic shell can hold a maximum of eight electrons, and the third atomic shell can house a maximum of eight electrons. The negatively charged electrons orbiting the nucleus are attracted to the positively charged protons in the nucleus via electromagnetic force. The attraction of opposite electrical charges gives rise to chemical bonds, which refers to the ways atoms are attached to each other.

The **atomic number** of an atom is determined by the number of protons within the nucleus. When a substance is composed of atoms that all have the same atomic number, it is called an **element**. Elements are arranged by atomic number and grouped by properties in the **Periodic table.**

An atom's **mass number** is determined by the sum of the total number of protons and neutrons in the atom. Most nuclei have a net neutral charge, and all atoms of one type have the same atomic number. However, there are some atoms of the same type that have a different mass number, due to an imbalance of neutrons. These are called **isotopes**. In isotopes, the atomic number, which is determined by the number of protons, is the same, but the mass number, which is determined by adding the protons and neutrons, is different due to the irregular number of neutrons.

Chemical Bonding

Chemical bonding typically results in the formation of a new substance, called a compound. Only the electrons in the outermost atomic shell are able to form chemical bonds. These electrons are known as **valence electrons**, and they are what determines the chemical properties of an atom.

Chemical bonding occurs between two or more atoms that are joined together. There are three types of chemical bonds: ionic, covalent, and metallic. The characteristics of the different bonds are determined by how electrons behave in a compound. **Lewis structures** were developed to help visualize the electrons in molecules; they are a method of writing a compound structure formula and including its electron composition. A Lewis symbol for an element consists of the element symbol and a dot for each valence electron. The dots are located on all four sides of the symbol, with a maximum of two dots per side, and eight dots, or electrons, total. The octet rule states that atoms tend to gain, lose, or share electrons until they have a total of eight valence electrons.

Ionic bonds are formed from the electrostatic attractions between oppositely charged atoms. They result from the transfer of electrons from a metal on the left side of the periodic table to a nonmetal on the right side. The metallic substance often has low ionization energy and will transfer an electron easily to the nonmetal, which has a high electron affinity. An example of this is the compound NaCl, which is sodium chloride or table salt, where the Na atom transfers an electron to the Cl atom. Due to strong bonding, ionic compounds have several distinct characteristics. They have high melting and boiling points and are brittle and crystalline. They are arranged in rigid, well-defined structures, which allow them to break apart along smooth, flat surfaces. The formation of ionic bonds is a reaction that is exothermic. In the opposite scenario, the energy it takes to break up a one mole quantity of an ionic compound is referred to as lattice energy, which is generally endothermic. The Lewis structure for NaCl is written as follows:

$$Na\cdot + :\ddot{\underset{\cdot\cdot}{Cl}}\cdot \longrightarrow Na^{+} + :\ddot{\underset{\cdot\cdot}{Cl}}^{-}:$$

Covalent bonds are formed when two atoms share electrons, instead of transferring them as in ionic compounds. The atoms in covalent compounds have a balance of attraction and repulsion between their protons and electrons, which keeps them bonded together. Two atoms can be joined by single, double, or even triple covalent bonds. As the number of electrons that are shared increases, the length of the bond decreases. Covalent substances have low melting and boiling points and are poor conductors of heat and electricity.

The Lewis structure for Cl_2 is written as follows:

Lewis structure Cl_2

$$:\ddot{\underset{\cdot\cdot}{Cl}}\cdot + \cdot\ddot{\underset{\cdot\cdot}{Cl}}: \longrightarrow :\ddot{\underset{\cdot\cdot}{Cl}}:\ddot{\underset{\cdot\cdot}{Cl}}:$$

Metallic bonds are formed by electrons that move freely through metal. They are the product of the force of attraction between electrons and metal ions. The electrons are shared by many metal cations and act like glue that holds the metallic substance together, similar to the attraction between oppositely-charged atoms in ionic substances, except the electrons are more fluid and float around the bonded metals and form a sea of electrons. Metallic compounds have characteristic properties that include strength, conduction of heat and electricity, and malleability. They can conduct electricity by passing energy through the freely moving electrons, creating a **current**. These compounds also have high melting and boiling points. Lewis structures are not common for metallic structures because of the free-roaming ability of the electrons.

Periodic Table

The periodic table catalogues all of the elements known to man, currently 118. It is one of the most important references in the science of chemistry. Information that can be gathered from the periodic table includes the element's atomic number, atomic mass, and chemical symbol. The first periodic table was rendered by Mendeleev in the mid-1800s and was ordered according to increasing atomic mass. The modern periodic table is arranged in order of increasing atomic number. It is also arranged in horizontal rows known as **periods,** and vertical columns known as **families,** or **groups**. The periodic table contains seven periods and eighteen families. Elements in the periodic table can also be classified into three major groups: metals, metalloids, and nonmetals. **Metals** are concentrated on the left side of the periodic table, while **nonmetals** are found on the right side. **Metalloids** occupy the area between the metals and nonmetals.

Due to the fact the periodic table is ordered by increasing atomic number, the electron configurations of the elements show periodicity. As the atomic number increases, electrons gradually fill the shells of an atom. In general, the start of a new period corresponds to the first time an electron inhabits a new shell.

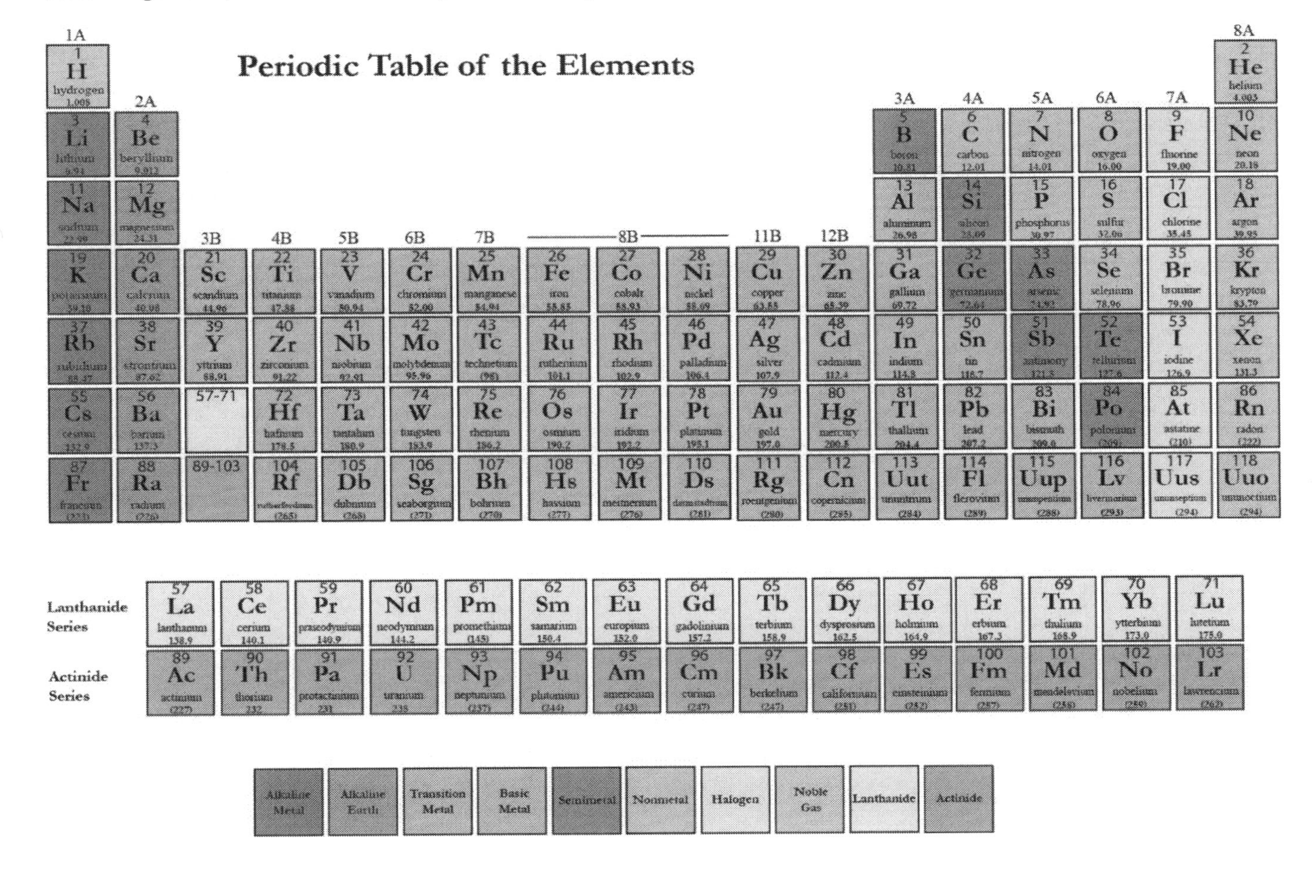

Other trends in the properties of elements in the periodic table are:

Atomic radius: One-half the distance between the nuclei of atoms of the same element.

Electronegativity: A measurement of the willingness of an atom to form a chemical bond.

Ionization energy: The amount of energy needed to remove an electron from a gas or ion.

Electron affinity: The ability of an atom to accept an electron.

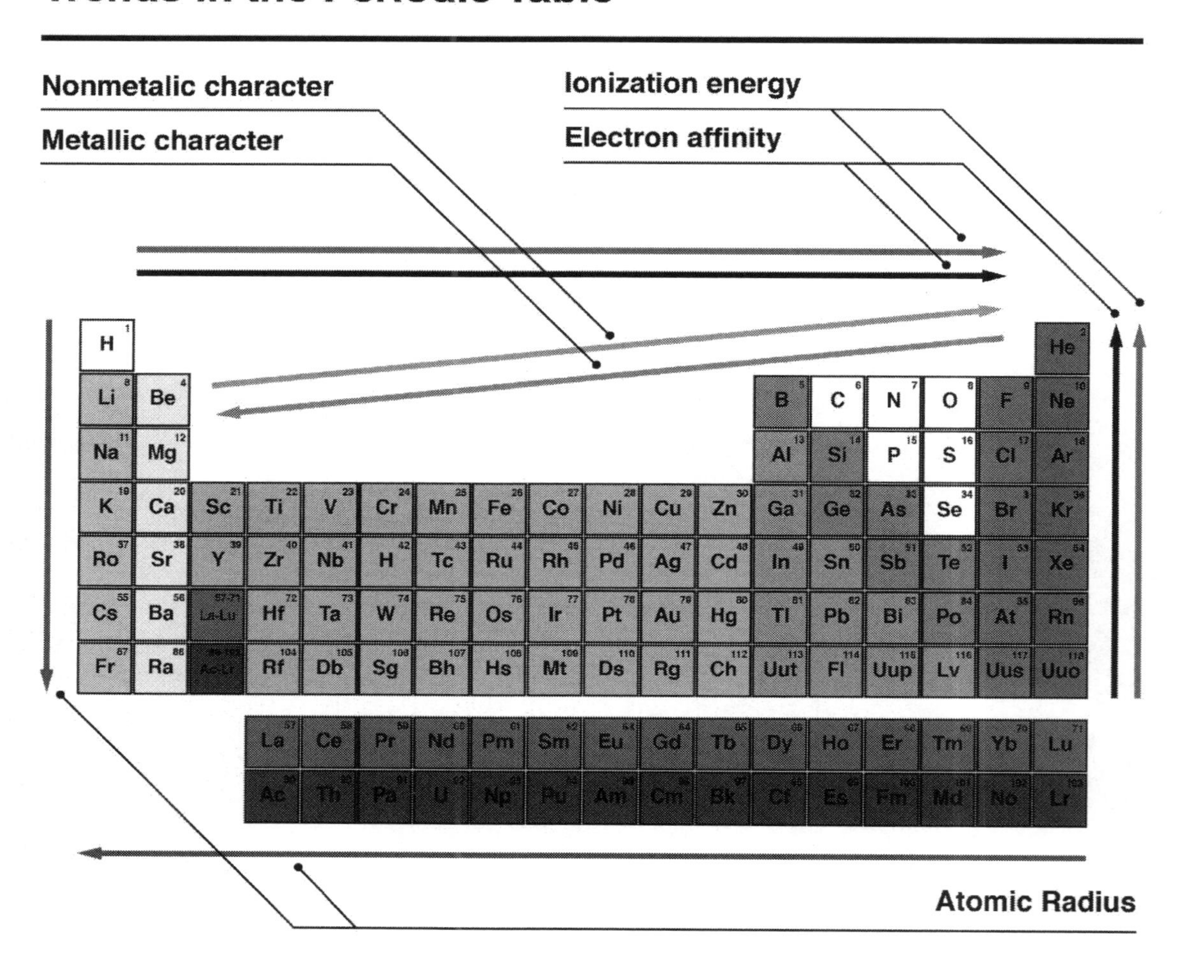

Chemical Equations

Chemical reactions are represented by **chemical equations**. The equations help to explain how the molecules change during the reaction. For example, when hydrogen gas (H_2) combines with oxygen gas (O_2), two molecules of water are formed. The equation is written as follows, where the "+" sign means *reacts with* and the "→" means *produces*:

$$2\,H_2 + O_2 \rightarrow 2\,H_2O$$

Two hydrogen molecules react with an oxygen molecule to produce two water molecules. In all chemical equations, the quantity of each element on the reactant side of the equation should equal the quantity of the same element on the product side of the equation due to the law of conservation of matter. If this is true, the equation is described as balanced. To figure out how many of each element there is on each side of the equation, the coefficient of the element should be multiplied by the subscript next to the element. Coefficients and subscripts are noted for quantities larger than one. The **coefficient** is the number located directly to the left of the element. The **subscript** is the small-sized number directly to the right of the element. In the equation above, on the left side, the coefficient of the hydrogen is two and the subscript is also two, which makes a total of four hydrogen atoms. Using the same method, there are two oxygen atoms. On the right side, the coefficient two is multiplied by the subscript in each element of the water molecule, making four hydrogen atoms and two oxygen atoms. This equation is balanced because there are four hydrogen atoms and two oxygen atoms on each side. The states of the reactants and products can also be written in the equation: gas (g), liquid (l), solid (s), and dissolved in water (aq). If they are included, they are noted in parentheses on the right side of each molecule in the equation.

Reaction Rates, Equilibrium, and Reversibility

Chemical reactions are conveyed using chemical equations. Chemical equations must be balanced with equivalent numbers of atoms for each type of element on each side of the equation. Antoine Lavoisier, a French chemist, was the first to propose the **Law of Conservation of Mass** for the purpose of balancing a chemical equation. The law states, "Matter is neither created nor destroyed during a chemical reaction."

The **reactants** are located on the left side of the arrow, while the **products** are located on the right side of the arrow. Coefficients are the numbers in front of the chemical formulas. Subscripts are the numbers to the lower right of chemical symbols in a formula. To tally atoms, one should multiply the formula's coefficient by the subscript of each chemical symbol. For example, the chemical equation $2\ H_2 + O_2 \rightarrow 2H_2O$ is balanced. For H, the coefficient of 2 multiplied by the subscript 2 = 4 hydrogen atoms. For O, the coefficient of 1 multiplied by the subscript 2 = 2 oxygen atoms. Coefficients and subscripts of 1 are understood and never written.

Reaction Rates

The rate of a reaction is the measure of the change in concentration of the reactants or products over a certain period of time. Many factors affect how fast or slow a reaction occurs, such as concentration, pressure, or temperature. As the concentration of a reactant increases, the rate of the reaction also increases, because the frequency of collisions between elements increases. High-pressure situations for reactants that are gases cause the gas to compress and increase the frequency of gas molecule collisions, similar to solutions with higher concentrations. Reactions rates are then increased with the higher frequency of gas molecule collisions. Higher temperatures usually increase the rate of the reaction, adding more energy to the system with heat and increasing the frequency of molecular collisions.

Equilibrium

Equilibrium is described as the state of a system when no net changes occur. Chemical equilibrium occurs when opposing reactions occur at equal rates. In other words, the rate of reactants forming products is equal to the rate of the products breaking down into the reactants—the concentration of reactants and products in the system doesn't change. This happens in **reversible chemical reactions** as opposed to irreversible chemical reactions. In **irreversible chemical reactions**, the products cannot be changed back to reactants. Although the concentrations are not changing in equilibrium, the forward and reverse reactions are likely still occurring. This type of equilibrium is called a **dynamic equilibrium**. In situations where all reactions have ceased, a **static equilibrium** is reached. Chemical equilibriums are also described as homogeneous or heterogeneous. **Homogeneous equilibrium** involves substances that are all in the same phase, while **heterogeneous equilibrium** means the substances are in different phases when equilibrium is reached.

When a reaction reaches equilibrium, the conditions of the equilibrium are described by the following equation, based on the chemical equation $aA + bB \leftrightarrow cC + dD$:

Catalysts are substances that accelerate the speed of a chemical reaction. A catalyst remains unchanged throughout the course of a chemical reaction. In most cases, only small amounts of a catalyst are needed. Catalysts increase the rate of a chemical reaction by providing an alternate path requiring less activation energy. Activation energy refers to the amount of energy required for the initiation of a chemical reaction.

Catalysts can be homogeneous or heterogeneous. Catalysts in the same phase of matter as its reactants are homogeneous, while catalysts in a different phase than reactants are heterogeneous. It is important to remember catalysts are selective. They don't accelerate the speed of all chemical reactions, but catalysts do accelerate specific chemical reactions.

Solutions and Solution Concentrations

A homogeneous mixture, also called a **solution,** has uniform properties throughout a given sample. An example of a homogeneous solution is salt fully dissolved in warm water. In this case, any number of samples taken from the parent solution would be identical.

One **mole** is the amount of matter contained in 6.02×10^{23} of any object, such as atoms, ions, or molecules. It is a useful unit of measure for items in large quantities. This number is also known as **Avogadro's number**. One mole of ^{12}C atoms is equivalent to 6.02×10^{23} ^{12}C atoms. Avogadro's number is often written as an inverse mole, or as 6.02×10^{23}/mol.

Molarity is the concentration of a solution. It is based on the number of moles of solute in one liter of solution and is written as the capital letter M. A 1.0 molar solution, or 1.0 M solution, has one mole of solute per liter of solution. The molarity of a solution can be determined by calculating the number of moles of the solute and dividing it by the volume of the solution in liters. The resulting number is the mol/L or M for molarity of the solution.

Chemical Reactions

Chemical reactions are characterized by a chemical change in which the starting substances, or reactants, differ from the substances formed, or products. Chemical reactions may involve a change in color, the production of gas, the formation of a precipitate, or changes in heat content.

The following are the five basic types of chemical reactions:

- **Decomposition Reactions:** A compound is broken down into smaller elements. For example, $2H_2O \rightarrow 2H_2 + O_2$. This is read as, "2 molecules of water decompose into 2 molecules of hydrogen and 1 molecule of oxygen."
- **Synthesis Reactions:** Two or more elements or compounds are joined together. For example, $2H_2 + O_2 \rightarrow 2H_2O$. This is read as, "2 molecules of hydrogen react with 1 molecule of oxygen to produce 2 molecules of water."
- **Single Displacement Reactions:** A single element or ion takes the place of another element in a compound. It is also known as a substitution reaction. For example, $Zn + 2\ HCl \rightarrow ZnCl_2 + H_2$. This is read as, "zinc reacts with 2 molecules of hydrochloric acid to produce one molecule of zinc chloride and one molecule of hydrogen." In other words, zinc replaces the hydrogen in hydrochloric acid.
- **Double Displacement Reactions:** Two elements or ions exchange a single element to form two different compounds, resulting in different combinations of cations and anions in the final compounds. It is also known as a metathesis reaction. For example, $H_2SO_4 + 2\ NaOH \rightarrow Na_2\ SO_4 + 2\ H_2O$
    - Special types of double displacement reactions include:
        - **Oxidation-Reduction (or Redox) Reactions:** Elements undergo a change in oxidation number. For example, $2\ S_2O_3^{2-}\ (aq) + I_2\ (aq) \rightarrow S_4O_6^{2-}\ (aq) + 2\ I^-\ (aq)$.
        - **Acid-Base Reactions:** Involves a reaction between an acid and a base, which produces a salt and water. For example, $HBr + NaOH \rightarrow NaBr + H_2O$.
        - **Combustion Reactions:** A hydrocarbon (a compound composed of only hydrogen and carbon) reacts with oxygen (O_2) to form carbon dioxide (CO_2) and water (H_2O). For example, $CH_4 + 2O_2 \rightarrow CO_2 + 2H_2O$.

Stoichiometry

Stoichiometry investigates the quantities of chemicals that are consumed and produced in chemical reactions. Chemical equations are made up of reactants and products; stoichiometry helps elucidate how the changes from reactants to products occur, as well as how to ensure the equation is balanced.

Chemical reactions are limited by the amount of starting material, or reactants, available to drive the process forward. The reactant that has the smallest amount of substance is called the limiting reactant. The **limiting reactant** is completely consumed by the end of the reaction. The other reactants are called **excess reactants**. For example, gasoline is used in a combustion reaction to make a car move and is the limiting reactant of the reaction. If the gasoline runs out, the combustion reaction can no longer take place, and the car stops.

The quantity of product that should be produced after using up all of the limiting reactant can be calculated and is called the **theoretical yield of the reaction**. Since the reactants do not always act as they should, the actual amount of resulting product is called the **actual yield**. The actual yield is divided by the theoretical yield and then multiplied by 100 to find the **percent yield** for the reaction.

Solution stoichiometry deals with quantities of solutes in chemical reactions that occur in solutions. The quantity of a solute in a solution can be calculated by multiplying the molarity of the solution by the volume. Similar to chemical equations involving simple elements, the number of moles of the elements that make up the solute should be equivalent on both sides of the equation.

When the concentration of a particular solute in a solution is unknown, a **titration** is used to determine that concentration. In a titration, the solution with the unknown solute is combined with a standard solution, which is a solution with a known solute concentration. The point at which the unknown solute has completely reacted with the known solute is called the **equivalence point**. Using the known information about the standard solution, including the concentration and volume, and the volume of the unknown solution, the concentration of the unknown solute is determined in a balanced equation. For example, in the case of combining acids and bases, the equivalence point is reached when the resulting solution is neutral. HCl, an acid, combines with NaOH, a base, to form water, which is neutral, and a solution of Cl^- ions and Na^+ ions. Before the equivalence point, there are an unequal number of cations and anions and the solution is not neutral.

Oxidation and Reduction

Oxidation and reduction reactions, also known as **redox reactions**, are those in which electrons are transferred from one element to another. Batteries and fuel cells are two energy-related technologies that utilize these reactions. When an atom, ion, or molecule loses its electrons and becomes more positively charged, it is described as being oxidized. When a substance gains electrons and becomes more negatively charged, it is reduced. In chemical reactions, if one element or molecule is oxidized, another must be reduced for the equation to be balanced. Although the transfer of electrons is obvious in some reactions where ions are formed, redox reactions also include those in which electrons are transferred but the products remain neutral.

Keep track of oxidation states or oxidation numbers to ensure the chemical equation is balanced. **Oxidation numbers** are assigned to each atom in a neutral substance or ion. For ions made up of a single atom, the oxidation number is equal to the charge of the ion. For atoms in their original elemental form, the oxidation number is always zero. Each hydrogen atom in an H_2 molecule, for example, has an oxidation number of zero. The sum of the oxidation numbers in a molecule should be equal to the overall charge of the molecule. If the molecule is a positively charged ion, the sum of the oxidation number should be equal to overall positive charge of the molecule. In ionic compounds that have a cation and anion joined, the sum of the oxidation numbers should equal zero.

All chemical equations must have the same number of elements on each side of the equation to be balanced. Redox reactions have an extra step of counting the electrons on both sides of the equation to be balanced. Separating redox reactions into oxidation reactions and reduction reactions is a simple way to account for all of the electrons involved. The individual equations are known as **half-reactions**. The number of electrons lost in the oxidation reaction must be equal to the number of electrons gained in the reduction reaction for the redox reaction to be balanced.

The oxidation of tin (Sn) by iron (Fe) can be balanced by the following half-reactions:

Oxidation: $Sn^{2+} \rightarrow Sn^{4+} + 2e^-$

Reduction: $2Fe^{3+} + 2e^- \rightarrow 2Fe^{2+}$

Complete redox reaction: $Sn^{2+} + 2Fe^{3+} \rightarrow Sn^{4+} + 2Fe^{2+}$

Acids and Bases

Acids and bases are defined in many different ways. An **acid** can be described as a substance that increases the concentration of H^+ ions when it is dissolved in water, as a proton donor in a chemical equation, or as an electron-pair acceptor. A **base** can be a substance that increases the concentration of OH^- ions when it is dissolved in water, accepts a proton in a chemical reaction, or is an electron-pair donor.

pH refers to the power or potential of hydrogen atoms and is used as a scale for a substance's acidity. In chemistry, pH represents the hydrogen ion concentration (written as $[H^+]$) in an aqueous, or watery, solution. The hydrogen ion concentration, $[H^+]$, is measured in moles of H^+ per liter of solution.

The pH scale is a logarithmic scale used to quantify how acidic or basic a substance is. pH is the negative logarithm of the hydrogen ion concentration: $pH = -\log [H^+]$. A one-unit change in pH correlates with a ten-fold change in hydrogen ion concentration. The pH scale typically ranges from zero to 14, although it is possible to have pHs outside of this range. Pure water has a pH of 7, which is considered **neutral**. pH values less than 7 are considered **acidic**, while pH values greater than 7 are considered **basic**, or **alkaline**.

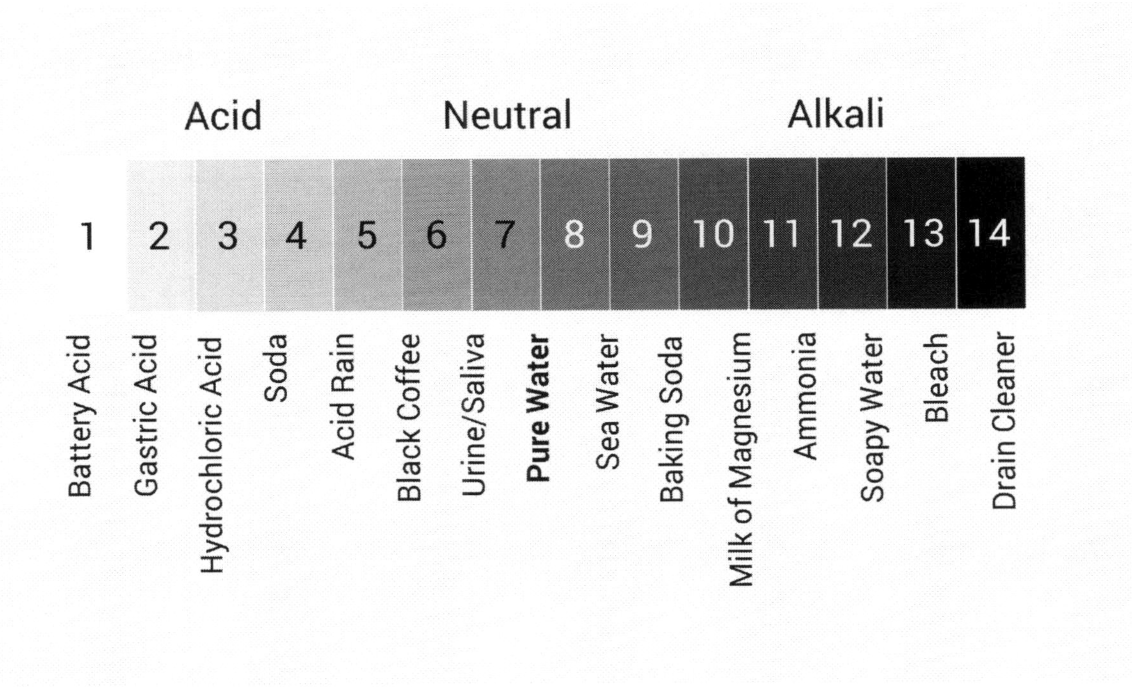

Generally speaking, an acid is a substance capable of donating hydrogen ions, while a base is a substance capable of accepting hydrogen ions. A **buffer** is a molecule that can act as either a hydrogen ion donor or acceptor. Buffers are crucial in the blood and body fluids and prevent the body's pH from fluctuating into dangerous territory. pH can be measured using a pH meter, test paper, or indicator sticks.

Water can act as either an acid or a base. When mixed with an acid, water can accept a proton and become an H_3O^+ ion. When mixed with a base, water can donate a proton and become an OH^- ion. Sometimes water molecules donate and accept protons from each other; this process is called **autoionization**. The chemical equation is written as follows: $H_2O + H_2O \rightarrow OH^- + H_3O^+$.

Acids and bases are characterized as strong, weak, or somewhere in between. Strong acids and bases completely or almost completely ionize in aqueous solution. The chemical reaction is driven completely forward, to the right side of the equation, where the acidic or basic ions are formed. Weak acids and bases do not completely disassociate in aqueous solution. They only partially ionize and the solution becomes a mixture of the acid or base, water, and the acidic or basic ions. Strong acids are complemented by weak bases, and vice versa. A **conjugate acid** is an ion that forms when its base pair gains a proton. For example, the conjugate acid NH_4^+ is formed from the base NH_3. The **conjugate base** that pairs with an acid is the ion that is formed when an acid loses a proton. NO_2^- is the conjugate base of the acid HNO_2.

Earth/Space Sciences

Types and Basic Characteristics of Rocks and Minerals and Their Formation Processes

The Rock Cycle

Although it may not always be apparent, rocks are constantly being destroyed while new ones are created in a process called the **rock cycle**. This cycle is driven by plate tectonics and the water cycle, which are discussed in detail later. The rock cycle starts with **magma**, the molten rock found deep within the Earth. As magma moves toward the Earth's surface, it hardens and transforms into igneous rock. Then, over time, igneous rock is broken down into tiny pieces called **sediment** that are eventually deposited all over the surface. As more and more sediment accumulates, the weight of the newer sediment compresses the older sediment underneath and creates sedimentary rock. As sedimentary rock is pushed deeper below the surface, the high pressure and temperature transform it into metamorphic rock. This metamorphic rock can either rise to the surface again or sink even deeper and melt back into magma, thus starting the cycle again.

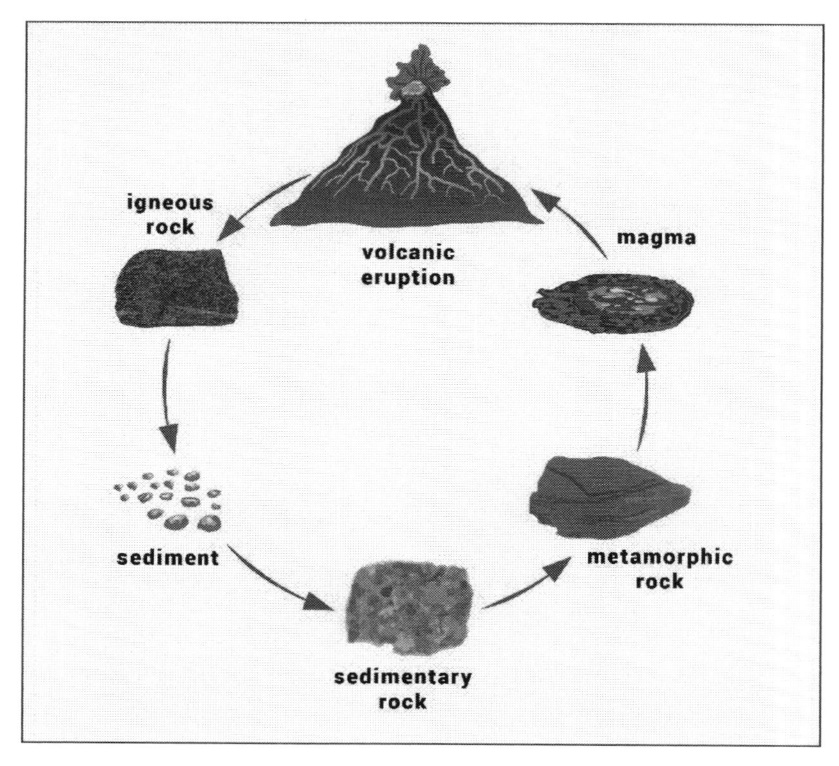

Characteristics of Rocks and Their Formation Processes

There are three main types of rocks: sedimentary, igneous, and metamorphic. Aside from physical characteristics, one of their main differences is how they are created. **Sedimentary rocks** are formed at the surface, on land and in bodies of water, through processes called deposition and cementation. They can be classified as clastic, biochemical, and chemical. **Clastic rocks**, such as sandstone, are composed of other pieces of inorganic rocks and sediment. **Biochemical rocks** are created from an organic material, such as coal, forming from dead plant life. **Chemical rocks** are created from the deposition of dissolved minerals, such as calcium salts that form stalagmites and stalactites in caves.

Igneous rocks are created when magma solidifies at or near the Earth's surface. When they're formed at the surface, (i.e. from volcanic eruption), they are **extrusive**. When they form below the surface, they're called **intrusive**. Examples of extrusive rocks are obsidian and tuff, while rocks like granite are intrusive.

Metamorphic rocks are the result of a transformation from other rocks. Based on appearance, these rocks are classified as foliated or non-foliated. **Foliated rocks** are created from compression in one direction, making them appear layered or folded like slate. **Non-foliated rocks** are compressed from all directions, giving them a more homogenous appearance, such as marble.

Characteristics of Minerals and Their Formation Processes

A **mineral**, such as gold, is a naturally occurring inorganic solid composed of one type of molecule or element that's organized into a crystalline structure. Rocks are aggregates of different types of minerals. Depending on their composition, minerals can be mainly classified into one of the following eight groups:

- **Carbonates:** formed from molecules that have either a carbon, nitrogen, or boron atom at the center.
- **Elements:** formed from single elements that occur naturally; includes metals such as gold and nickel, as well as metallic alloys like brass.
- **Halides:** formed from molecules that have halogens; halite, which is table salt, is a classic example.
- **Oxides:** formed from molecules that contain oxygen or hydroxide and are held together with ionic bonds; encompasses the phosphates, silicates, and sulfates.
- **Phosphates:** formed from molecules that contain phosphates; the apatite group minerals are in this class.
- **Silicates:** formed from molecules that contain silicon, silicates are the largest class and usually the most complex minerals; topaz is an example of a silicate.
- **Sulfates:** formed from molecules that contain either sulfur, chromium, tungsten, selenium, tellurium, and/or molybdenum.
- **Sulfides:** formed from molecules that contain sulfide (S^{2-}); includes many of the important metal ores, such as lead and silver.

One important physical characteristic of a mineral is its **hardness**, which is defined as its resistance to scratching. When two crystals are struck together, the harder crystal will scratch the softer crystal. The most common measure of hardness is the **Mohs Hardness Scale**, which ranges from 1 to 10, with 10 being the hardest. Diamonds are rated 10 on the Mohs Hardness Scale, and talc, which was once used to make baby powder, is rated 1. Other important characteristics of minerals include **luster** or shine, **color**, and **cleavage**, which is the natural plane of weakness at which a specific crystal breaks.

Erosion, Weathering, and Deposition of Earth's Surface Materials and Soil Formation

Erosion and Deposition

Erosion is the process of moving rock and occurs when rock and sediment are picked up and transported. Wind, water, and ice are the primary factors for erosion. **Deposition** occurs when the particles stop moving and settle onto a surface, which can happen through gravity or involve processes such as precipitation or flocculation. **Precipitation** is the solidification or crystallization of dissolved ions that occurs when a solution is oversaturated. **Flocculation** is similar to coagulation and occurs when colloid materials (materials that aren't dissolved but are suspended in the medium) aggregate or clump until they are too heavy to remain suspended.

Chemical and Physical (Mechanical) Weathering

Weathering is the process of breaking down rocks through mechanical or chemical changes. **Mechanical forces** include animal contact, wind, extreme weather, and the water cycle. These physical forces don't alter the composition of rocks. In contrast, chemical weathering transforms rock composition. When water and minerals interact, they can start chemical reactions and form new or secondary minerals from the original rock. In chemical weathering, the processes of oxidation and hydrolysis are important. When rain falls, it dissolves atmospheric carbon dioxide and becomes acidic. With sulfur dioxide and nitrogen oxide in the atmosphere from volcanic eruptions or burning fossil fuels, the rainfall becomes even more acidic and creates **acid rain**. Acidic rain can dissolve the rock that it falls upon.

Characteristics of Soil

Soil is a combination of minerals, organic materials, liquids, and gases. There are three main types of soil, as defined by their compositions, going from coarse to fine: sand, silt, and clay. Large particles, such as those found in sand, affect how water moves through the soil, while tiny clay particles can be chemically active and bind with water and nutrients. An important characteristic of soil is its ability to form a crust when dehydrated. In general, the finer the soil, the harder the crust, which is why clay (and not sand) is used to make pottery.

There are many different classes of soil, but the components are always sand, silt, or clay. Below is a chart used by the United States Department of Agriculture (USDA) to define soil types:

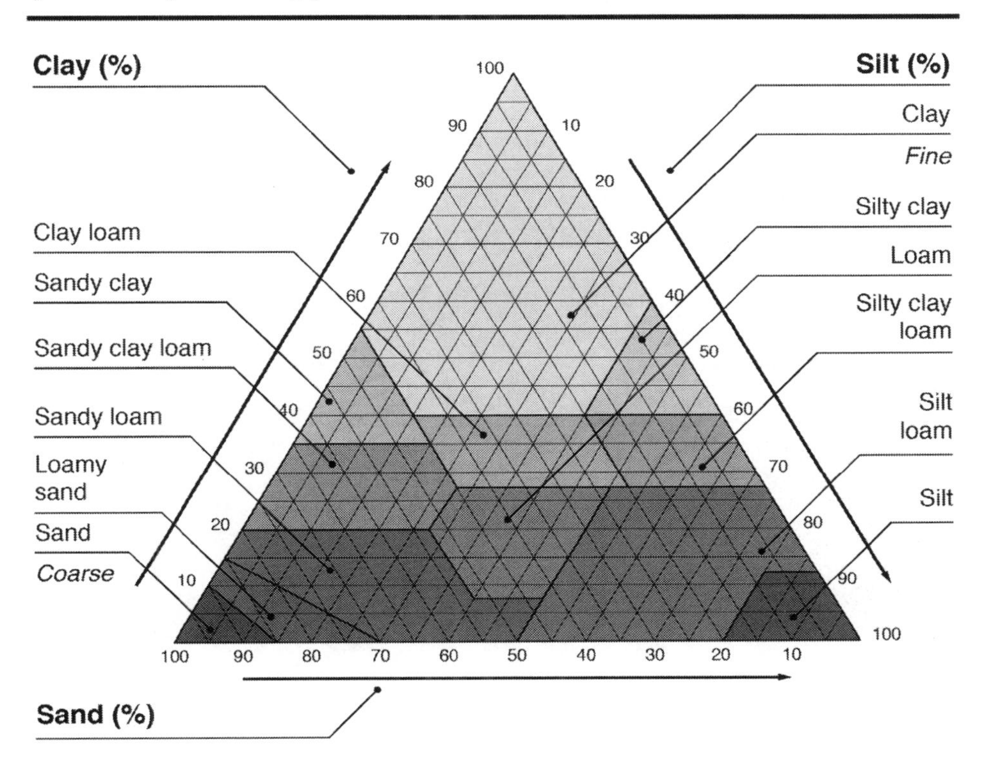

Loam is a term for soil that is a mixture of sand, silt, and clay. It's also the soil most commonly used for agriculture and gardening.

Porosity and Permeability

Porosity and permeability refer to how water moves through rock and soil underground. **Porosity** is a measure of the open space in a rock. This space can be between grains or within cracks and cavities in the rock. **Permeability** is a measure of the ease with which water can move through a porous rock. Therefore, rock that's more porous is also more permeable. When a rock is more permeable, it's less effective as a water purifier because dirty particles in the water can pass through porous rock.

Runoff and Infiltration

An important function of soil is to absorb water to be used by plants or released into groundwater. **Infiltration capacity** is the maximum amount of water that can enter soil at any given time and is regulated by the soil's porosity and composition. For example, sandy soils have larger pores than clays, allowing water to infiltrate them easier and faster. **Runoff** is water that moves across land's surface and may end up in a stream or a rut in the soil. Runoff generally occurs after the soil's infiltration capacity is reached. However, during heavy rainfalls, water may reach the soil's surface at a faster rate than infiltration can occur, causing runoff without soil saturation. In addition, if the ground is frozen and the soil's pores are blocked by ice, runoff may occur without water infiltrating the soil.

Earth's Basic Structure and Internal Processes

Earth's Layers

Earth has three major layers: a thin solid outer surface or **crust**, a dense **core**, and a **mantle** between them that contains most of the Earth's matter. This layout resembles an egg, where the eggshell is the crust, the mantle is the egg white, and the core is the yolk. The outer crust of the Earth consists of igneous or sedimentary rocks over metamorphic rocks. Together with the upper portion of the mantle, it forms the **lithosphere**, which is broken into tectonic plates.

Major plates of the lithosphere

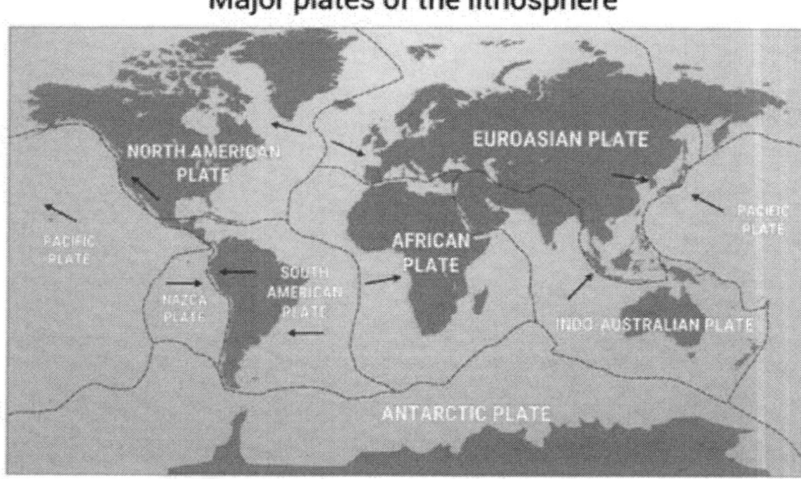

The mantle can be divided into three zones. The **upper mantle** is adjacent to the crust and composed of solid rock. Below the upper mantle is the **transition zone**. The **lower mantle** below the transition zone is a layer of completely solid rock. Underneath the mantle is the molten **outer core** followed by the compact, solid **inner core**. The inner and outer cores contain the densest elements, consisting of mostly iron and nickel.

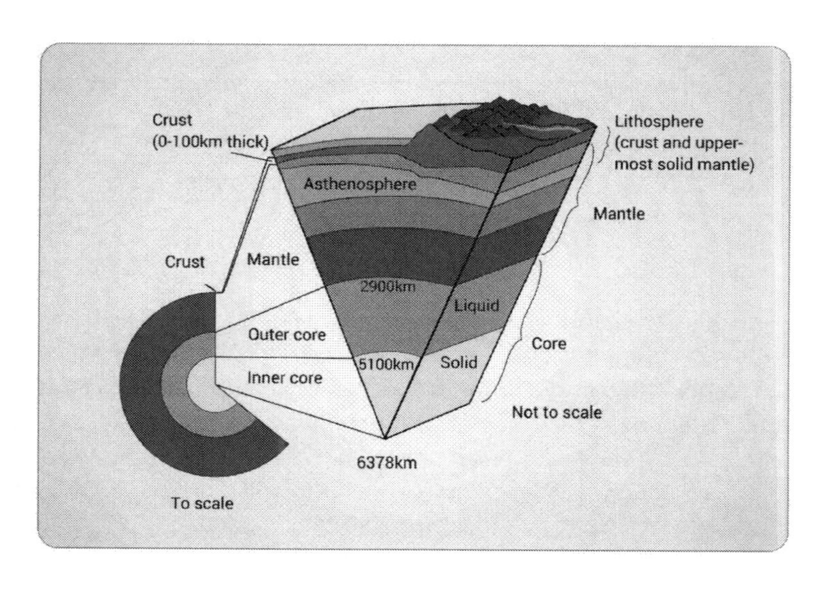

Shape and Size of the Earth

The Earth isn't a perfect sphere; it's slightly elliptical. From center to surface, its radius is almost 4,000 miles, and its circumference around the equator is about 24,902 miles. In comparison, the Sun's radius is 432,288 miles—over 1,000 times larger than the Earth's—and the Moon's radius is about 1,000 miles.

Geographical Features

The Earth's surface is dynamic and consists of various landforms. As tectonic plates are pushed together, **mountains** are formed. **Canyons** are deep trenches that are usually created by plates moving apart but can also be created by constant weathering and erosion from rivers and runoff. **Deltas** are flat, triangular stretches of land formed by rivers that deposit sediment and water into the ocean. **Sand dunes** are mountains of sand located in desert areas or the bottom of the ocean. They are formed by wind and water movement when there's an absence of plants or other features that would otherwise hold the sand in place.

The Earth's Magnetic Field

The Earth's **magnetic field** is created by the magnetic forces that extend from the Earth's interior to outer space. It can be modeled as a magnetic dipole tilted about 10 degrees from the Earth's rotational axis, as if a bar magnet was placed at an angle inside the Earth's core. The **geomagnetic pole** located near Greenland in the northern hemisphere is actually the south pole of the Earth's magnetic field, and vice versa for the southern geomagnetic pole. The **magnetosphere** is the Earth's magnetic field, which extends tens of thousands of kilometers into space and protects the Earth and the atmosphere from damaging solar wind and cosmic rays.

Plate Tectonics Theory and Evidence

The theory of **plate tectonics** hypothesizes that the continents weren't always separated like they are today but were once joined and slowly drifted apart. Evidence for this theory is based upon the fossil record. Fossils of one species were found in regions of the world now separated by an ocean. It's unlikely that a single species could have travelled across the ocean or that two separate species evolved into a single species.

Folding and Faulting

The exact number of tectonic plates is debatable, but scientists estimate there are around nine to fifteen major plates and almost 40 minor plates. The line where two plates meet is called a **fault.** The San Andreas Fault is where the Pacific and North American plates meet. Faults or boundaries are classified depending on the interaction between plates. Two plates collide at **convergent boundaries**. **Divergent boundaries** occur when two plates move away from each other. Tectonic plates can move vertically and horizontally.

Continental Drift, Seafloor Spreading, Magnetic Reversals

The movement of tectonic plates is similar to pieces of wood floating in a pool of water. They can bob up and down as well as bump, slide, and move away from each other. These different interactions create the Earth's landscape. The collision of plates can create mountain ranges, while their separation can create canyons or underwater chasms. One plate can also slide atop another and push it down into the Earth's hot mantle, creating magma and volcanoes, in a process called **subduction**.

Unlike a regular magnet, the Earth's magnetic field changes over time because it's generated by the motion of molten iron alloys in the outer core. Although the magnetic poles can wander geographically, they do so at such a slow rate that they don't affect the use of compasses in navigation. However, at irregular intervals that are several hundred thousand years long, the fields can reverse, with the north and south magnetic poles switching places.

Characteristics of Volcanoes

Volcanoes are mountainous structures that act as vents to release pressure and magma from the Earth's crust. During an **eruption**, the pressure and magma are released, and volcanoes smoke, rumble, and throw ash and **lava**, or molten rock, into the air. **Hot spots** are volcanic regions of the mantle that are hotter than surrounding regions.

Characteristics of Earthquakes

Earthquakes occur when tectonic plates slide or collide as a result of the crust suddenly releasing energy. Stress in the Earth's outer layer pushes together two faults. The motion of the planes of the fault continues until something makes them stop. The **epicenter** of an earthquake is the point on the surface directly above where the fault is slipping. If the epicenter is located under a body of water, the earthquake may cause a **tsunami**, a series of large, forceful waves.

Seismic waves and Triangulation

Earthquakes cause **seismic waves**, which travel through the Earth's layers and give out low-frequency acoustic energy. Triangulation of seismic waves helps scientists determine the origin of an earthquake.

The Water Cycle

Evaporation and Condensation

The **water cycle** is the cycling of water between its three physical states: solid, liquid, and gas. The Sun's thermal energy heats surface water so it evaporates. As water vapor collects in the atmosphere from evaporation, it eventually reaches a saturation level where it condenses and forms clouds heavy with water droplets.

Precipitation

When the droplets condense as clouds get heavy, they fall as different forms of precipitation, such as rain, snow, hail, fog, and sleet. **Advection** is the process of evaporated water moving from the ocean and falling over land as precipitation.

Runoff and Infiltration

Runoff and infiltration are important parts of the water cycle because they provide water on the surface available for evaporation. **Runoff** can add water to oceans and aid in the advection process. **Infiltration** provides water to plants and aids in the transpiration process.

Transpiration

Transpiration is an evaporation-like process that occurs in plants and soil. Water from the stomata of plants and from pores in soil evaporates into water vapor and enters the atmosphere.

Historical Geology

Principle of Uniformitarianism

Uniformitarianism is the assumption that natural laws and processes haven't changed and apply everywhere in the universe. In geology, uniformitarianism includes the **gradualist model**, which states that "the present is the key to the past" and claims that natural laws functioned at the same rates as observed today.

Basic Principles of Relative Age Dating

Relative age dating is the determination of the relative order of past events without determining absolute age. The Law of Superposition states that older geological layers are deeper than more recent layers. Rocks and fossils can be used to compare one stratigraphic column with another. A **stratigraphic column** is a drawing that describes the vertical location of rocks in a cliff wall or underground. Correlating these columns from different geographic areas allows scientists to understand the relationships between different areas and strata. Before the discovery of radiometric dating, geologists used this technique to determine the ages of different materials. Relative dating can only determine the sequential order of events, not the exact time they occurred. The **Law of Fossil Succession** states that when the same kinds of fossils are found in rocks from different locations, the rocks are likely the same age.

Absolute (Radiometric) Dating

Absolute or **radiometric dating** is the process of determining age on a specified chronology in archaeology and geology. It attempts to provide a numerical age by measuring the radioactive decay of elements (such as carbon-14) trapped in rocks or minerals and using the known rate of decay to determine how much time has passed.

Characteristics and Processes of the Earth's Oceans and Other Bodies of Water

Distribution and Location of the Earth's Water

A **body of water** is any accumulation of water on the Earth's surface. It usually refers to oceans, seas, and lakes, but also includes ponds, wetlands, and puddles. Rivers, streams, and canals are bodies of water that involve the movement of water.

Most bodies of water are naturally occurring geographical features, but they can also be artificially created like lakes created by dams. Saltwater oceans make up 96% of the water on the Earth's surface. Freshwater makes up 2.5% of the remaining water.

Seawater Composition

Seawater is water from a sea or ocean. On average, seawater has a salinity of about 3.5%, meaning every kilogram of seawater has approximately 35 grams of dissolved sodium chloride salt. The average density of saltwater at the surface is 1.025 kg/L, making it denser than pure or freshwater, which has a density of 1.00 kg/L. Because of the dissolved salts, the freezing point of saltwater is also lower than that of pure water; salt water freezes at −2 °C (28 °F). As the concentration of salt increases, the freezing point decreases. Thus, it's more difficult to freeze water from the Dead Sea—a saltwater lake known to have water with such high salinity that swimmers cannot sink.

Coastline Topography and the Topography of Ocean Floor

Topography is the study of natural and artificial features comprising the surface of an area. **Coastlines** are an intermediate area between dry land and the ocean floor. The ground progressively slopes from the dry coastal area to the deepest depth of the ocean floor. At the continental shelf, there's a steep descent of the ocean floor. Although it's often believed that the ocean floor is flat and sandy like a beach, its topography includes mountains, plateaus, and valleys.

Tides, Waves, and Currents

Tides are caused by the pull of the Moon and the Sun. When the Moon is closer in its orbit to the Earth, its gravity pulls the oceans away from the shore. When the distance between the Moon and the Earth is greater, the pull is weaker, and the water on Earth can spread across more land. This relationship creates low and high tides. Waves are influenced by changes in tides as well as the wind. The energy transferred from wind to the top of large bodies of water creates *crests* on the water's surface and waves below. Circular movements in the ocean are called **currents**. They result from the **Coriolis Effect**, which is caused by the Earth's rotation. Currents spin in a clockwise direction above the equator and counterclockwise below the equator.

Estuaries and Barrier Islands

An **estuary** is an area of water located on a coast where a river or stream meets the sea. It's a transitional area that's partially enclosed, has a mix of salty and fresh water, and has calmer water than the open sea. **Barrier islands** are coastal landforms created by waves and tidal action parallel to the mainland coast. They usually occur in chains, and they protect the coastlines and create areas of protected waters where wetlands may flourish.

Islands, Reefs, and Atolls

Islands are land that is completely surrounded by water. **Reefs** are bars of rocky, sandy, or coral material that sit below the surface of water. They may form from sand deposits or erosion of underwater rocks. An **atoll** is a coral reef in the shape of a ring (but not necessarily circular) that encircles a lagoon. In order for an atoll to exist, the rate of its erosion must be slower than the regrowth of the coral that composes the atoll.

Polar Ice, Icebergs, Glaciers

Polar ice is the term for the sheets of ice that cover the poles of a planet. **Icebergs** are large pieces of freshwater ice that break off from glaciers and float in the water. A **glacier** is a persistent body of dense ice that constantly moves because of its own weight. Glaciers form when snow accumulates at a faster rate than it melts over centuries. They form only on land, in contrast to **ice caps**, which can form from sheets of ice in the ocean. When glaciers deform and move due to stresses created by their own weight, they can create **crevasses** and other large distinguishing land features.

Lakes, Ponds, and Wetlands

Lakes and **ponds** are bodies of water that are surrounded by land. They aren't part of the ocean and don't contain flowing water. Lakes are larger than ponds, but otherwise the two bodies don't have a scientific distinction. **Wetlands** are areas of land saturated by water. They have a unique soil composition and provide a nutrient-dense area for vegetation and aquatic plant growth. They also play a role in water purification and flood control.

Streams, Rivers, and River Deltas

A **river** is a natural flowing waterway usually consisting of freshwater that flows toward an ocean, sea, lake, or another river. Some rivers flow into the ground and become dry instead of reaching another body of water. Small rivers are usually called **streams** or **creeks**. River **deltas** are areas of land formed from the sediment carried by a river and deposited before it enters another body of water. As the river reaches its end, the flow of water slows, and the river loses the power to transport the sediment so it falls out of suspension.

Geysers and Springs

A **spring** is a natural occurrence where water flows from an aquifer to the Earth's surface. A **geyser** is a spring that intermittently and turbulently discharges water. Geysers form only in certain hydrogeological conditions. They require proximity to a volcanic area or magma to provide enough heat to boil or vaporize the water. As hot water and steam accumulate, pressure grows and creates the spraying geyser effect.

Properties of Water that Affect Earth Systems

Water is a chemical compound composed of two hydrogen atoms and one oxygen atom (H_2O) and has many unique properties. In its solid state, water is less dense than its liquid form; therefore, ice floats in water. Water also has a very high **heat capacity**, allowing it to absorb a high amount of the Sun's energy without getting too hot or evaporating. Its chemical structure makes it a polar compound, meaning one side has a negative charge while the other is positive. This characteristic—along with its ability to form strong intermolecular hydrogen bonds with itself and other molecules—make water an effective solvent for other chemicals.

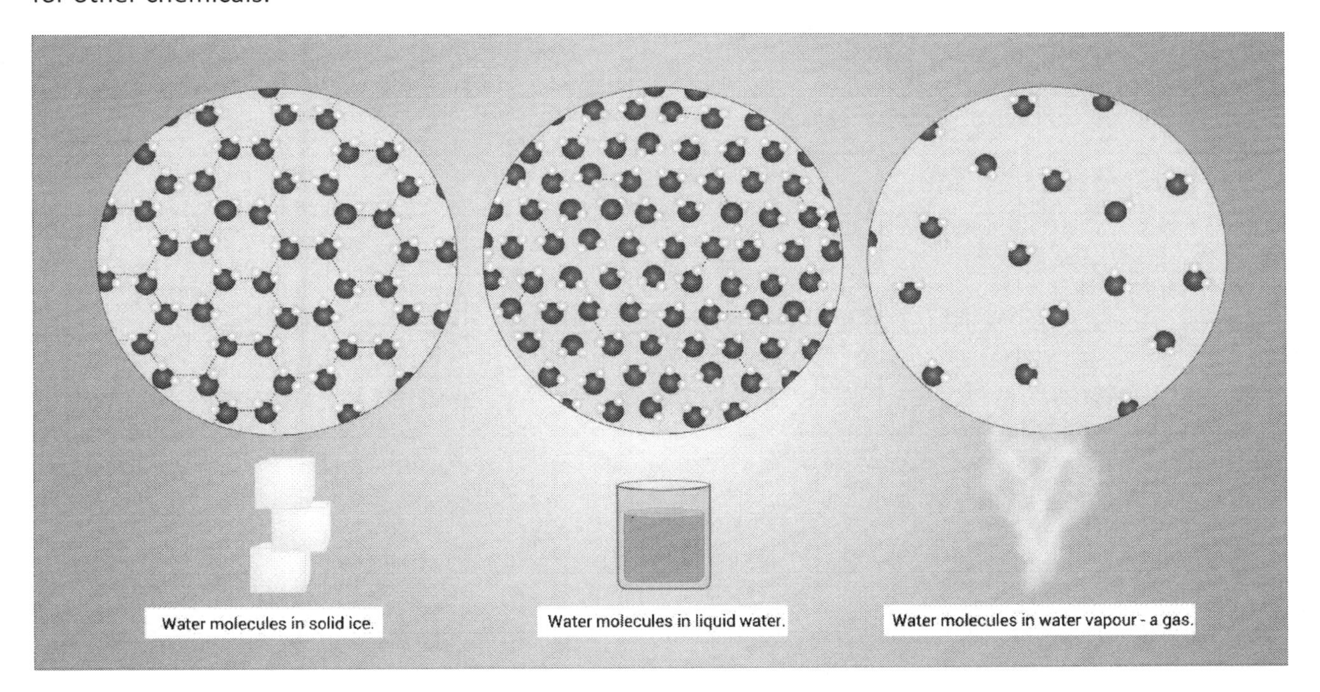

Basic Structure and Composition of the Earth's Atmosphere

Layers

The Earth's atmospheric layers are determined by their temperatures but are reported by their distance above sea level. Listed from closest to sea level on upward, the levels are:

- **Troposphere:** sea level to 11 miles above sea level
- **Stratosphere:** 11 miles to 31 miles above sea level
- **Mesosphere:** 31 miles to 50 miles above sea level
- **Ionosphere:** 50 miles to 400 miles above sea level
- **Exosphere***:* 400 miles to 800 miles above sea level

The ionosphere and exosphere are together considered the **thermosphere**. The ozone layer is in the stratosphere and weather experienced on Earth's surface is a product of factors in the troposphere.

Composition of the Atmosphere

The Earth's atmosphere is composed of gas particles: 78% nitrogen, 21% oxygen, 1% other gases such as argon, and 0.039% carbon dioxide. The atmospheric layers are created by the number of particles in the air and gravity's pull upon them.

Atmospheric Pressure and Temperature

The lower atmospheric levels have higher atmospheric pressures due to the mass of the gas particles located above. The air is less dense (it contains fewer particles per given volume) at higher altitudes. The temperature changes from the bottom to top of each atmospheric layer. The tops of the troposphere and mesosphere are colder than their bottoms, but the reverse is true for the stratosphere and thermosphere. Some of the warmest temperatures are actually found in the thermosphere because of a type of radiation that enters that layer.

Basic Concepts of Meteorology

Relative Humidity

Relative humidity is the ratio of the partial pressure of water vapor to water's equilibrium vapor pressure at a given temperature. At low temperatures, less water vapor is required to reach a high relative humidity. More water vapor is needed to reach a high relative humidity in warm air, which has a greater capacity for water vapor. At ground level or other areas of higher pressure, relative humidity increases as temperatures decrease because water vapor condenses as the temperature falls below the dew point. As relative humidity cannot be greater than 100%, the dew point temperature cannot be greater than the air temperature.

Dew Point

The **dew point** is the temperature at which the water vapor in air at constant barometric pressure condenses into liquid water due to saturation. At temperatures below the dew point, the rate of condensation will be greater than the rate of evaporation, forming more liquid water. When condensed water forms on a surface, it's called **dew**; when it forms in the air, it's called **fog** or **clouds**, depending on the altitude.

Wind

Wind is the movement of gas particles across the Earth's surface. Winds are generated by differences in atmospheric pressure. Air inherently moves from areas of higher pressure to lower pressure, which is what causes wind to occur. Surface friction from geological features, such as mountains or man-made features can decrease wind speed. In meteorology, winds are classified based on their strength, duration, and direction. **Gusts** are short bursts of high-speed wind, **squalls** are strong winds of intermediate duration (around one minute), and winds with a long duration are given names based on their average strength. **Breezes** are the weakest, followed by **gales**, **storms**, and **hurricanes**.

Cloud Types and Formation

Water in the atmosphere can exist as visible masses called *clouds* composed of water droplets, tiny crystals of ice, and various chemicals. Clouds exist primarily in the troposphere.

They can be classified based on the altitude at which they occur:

- **High-Clouds**—between 5,000 and 13,000 meters above sea level
    - Cirrus: thin and wispy "mare's tail" appearance
    - Cirrocumulus: rows of small puffy pillows
    - Cirrostratus: thin sheets that cover the sky
- **Middle clouds**—between 2,000 and 7,000 meters above sea level
    - Altocumulus: gray and white and made up of water droplets
    - Altostratus: grayish or bluish gray clouds
- **Low clouds**—below 2,000 meters above sea level
    - Stratus: gray clouds made of water droplets that can cover the sky
    - Stratocumulus: gray and lumpy low-lying clouds
    - Nimbostratus: dark gray with uneven bases; typical of rain or snow clouds

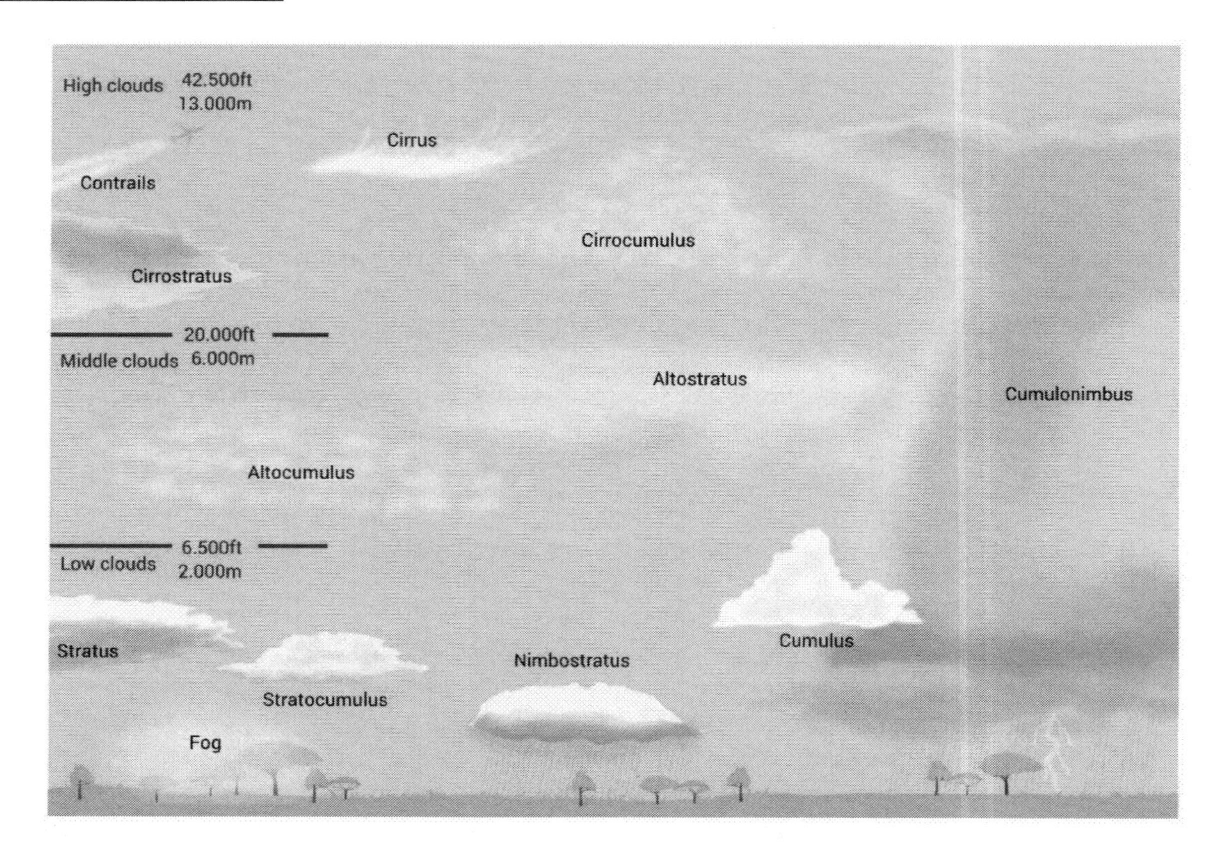

Types of Precipitation

There are three distinct processes by which precipitation occurs. **Convection precipitation** occurs when air rises vertically in a forceful manner, quickly overturning the atmosphere and resulting in heavy precipitation. It's generally more intense and shorter in duration than **stratiform precipitation**, which occurs when large masses of air move over each other. **Orographic precipitation** occurs when moist air is forced upwards over rising terrain, such as a mountain. Most storms are a result of convection precipitation.

Precipitation can fall in liquid or solid phases, as well as any form in between. Liquid precipitation includes rain and drizzle. Frozen precipitation includes snow, sleet, and hail. Intensity is classified by rate of fall or visibility restriction. The forms of precipitation are:

- **Rain:** water vapor that condenses on dust particles in the troposphere until it becomes heavy enough to fall to Earth
- **Sleet:** rain that freezes on its way down; it starts as ice that melts and then freezes again before hitting the ground
- **Hail:** balls of ice thrown up and down several times by turbulent winds, so that more and more water vapor can condense and freeze on the original ice; hail can be as large as golf balls or even baseballs
- **Snow:** loosely packed ice crystals that fall to Earth

Air Masses, Fronts, Storms, and Severe Weather

Air masses are volumes of air defined by their temperature and the amount of water vapor they contain. A **front** is where two air masses of different temperatures and water vapor content meet. Fronts can be the site of extreme weather, such as thunderstorms, which are caused by water particles rubbing against each other. When they do so, electrons are transferred and energy and electrical currents accumulate. When enough energy accumulates, thunder and lightning occur. **Lightning** is a massive electric spark created by a cloud, and **thunder** is the sound created by an expansion of air caused by the sudden increase in pressure and temperature around lightning.

Extreme weather includes tornadoes and hurricanes. **Tornadoes** are created by changing air pressure and winds that can exceed 300 miles per hour. **Hurricanes** occur when warm ocean water quickly evaporates and rises to a colder, low-pressure portion of the atmosphere. Hurricanes, typhoons, and tropical cyclones are all created by the same phenomena but they occur in different regions. **Blizzards** are similar to hurricanes in that they're created by the clash of warm and cold air, but they only occur when cold Arctic air moves toward warmer air. They usually involve large amounts of snow.

Development and Movement of Weather Patterns

A **weather pattern** is weather that's consistent for a period of time. Weather patterns are created by fronts. A **cold front** is created when two air masses collide in a high-pressure system. A **warm front** is created when a low-pressure system results from the collision of two air masses; they are usually warmer and less dense than high-pressure systems. When a cold front enters an area, the air from the warm front is forced upwards. The temperature of the warm front's air decreases, condenses, and often creates clouds and precipitation. When a warm front moves into an area, the warm air moves slowly upwards at an angle. Clouds and precipitation form, but the precipitation generally lasts longer because of how slowly the air moves.

Major Factors that Affect Climate and Seasons

Effects of Latitude, Geographical Location, and Elevation

The climate and seasons of different geographical areas are primarily dictated by their sunlight exposure. Because the Earth rotates on a tilted axis while travelling around the Sun, different latitudes get different amounts of direct sunlight throughout the year, creating different climates. Polar regions experience the greatest variation, with long periods of limited or no sunlight in the winter and up to 24 hours of daylight in the summer. Equatorial regions experience the least variance in direct sunlight exposure. Coastal areas experience breezes in the summer as cooler ocean air moves ashore, while areas southeast of the Great Lakes can get "lake effect" snow in the winter, as cold air travels over the warmer water and creates snow on land. Mountains are often seen with snow in the spring and fall. Their high elevation causes mountaintops to stay cold. The air around the mountaintop is also cold and holds less water vapor than air at sea level. As the water vapor condenses, it creates snow.

Effects of Atmospheric Circulation

Global winds are patterns of wind circulation and they have a major influence on global weather and climate. They help influence temperature and precipitation by carrying heat and water vapor around the Earth. These winds are driven by the uneven heating between the polar and equatorial regions created by the Sun. Cold air from the polar regions sinks and moves toward the equator, while the warm air from the equator rises and moves toward the poles. The other factor driving global winds is the **Coriolis**

Effect. As air moves from the North Pole to the equator, the Earth's rotation makes it seem as if the wind is also moving to the right, or westbound, and eastbound from South Pole to equator.

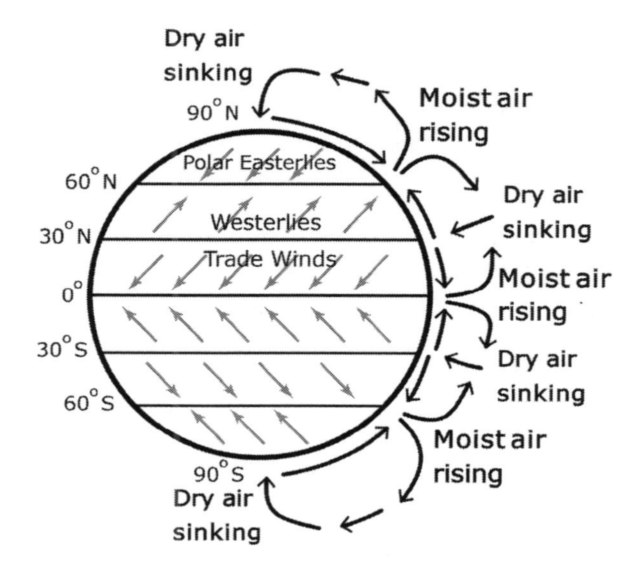

Global wind patterns are given names based on which direction they blow. There are three major wind patterns in each hemisphere. Notice the image above diagramming the movement of warm (dry) air and cold (moist) air.

Trade winds—easterly surface winds found in the troposphere near the equator—blow predominantly from the northeast in the Northern Hemisphere and from the southeast in the Southern Hemisphere. These winds direct the tropical storms that develop over the Atlantic, Pacific, and Indian Oceans and land in North America, Southeast Asia, and eastern Africa, respectively. **Jet streams** are westerly winds that follow a narrow, meandering path. The two strongest jet streams are the polar jets and the subtropical jets. In the Northern Hemisphere, the polar jet flows over the middle of North America, Europe, and Asia, while in the Southern Hemisphere, it circles Antarctica.

Effects of Ocean Circulation

Ocean currents are similar to global winds because winds influence how the oceans move. Ocean currents are created by warm water moving from the equator towards the poles while cold water travels from the poles to the equator. The warm water can increase precipitation in an area because it evaporates faster than the colder water.

Characteristics and Locations of Climate Zones

Climate zones are created by the Earth's tilt as it travels around the Sun. These zones are delineated by the equator and four other special latitudinal lines: the Tropic of Cancer or Northern Tropic at 23.5° North; the Tropic of Capricorn or Southern Tropic at 23.5° South; the Arctic Circle at 66.5° North; and the Antarctic Circle at 66.5° South. The areas between these lines of latitude represent different climate zones. **Tropical climates** are hot and wet, like rainforests, and tend to have abundant plant and animal life, while polar climates are cold and usually have little plant and animal life. **Temperate zones** can vary and experience the four seasons.

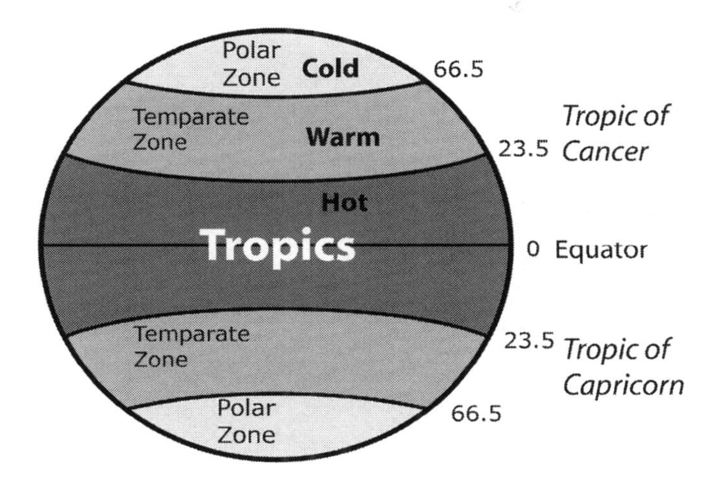

Effect of the Tilt of the Earth's Axis on Seasons

In addition to the equator and the prime meridian, other major lines of latitude and longitude divide the world into regions relative to the direct rays of the Sun. These lines correspond with the Earth's 23.5-degree tilt, and are responsible—along with the Earth's revolution around the Sun—for the seasons. For example, the Northern Hemisphere is tilted directly toward the Sun from June 22 to September 23, which creates the summer. Conversely, the Southern Hemisphere is tilted away from the Sun and experiences winter during those months. The area between the Tropic of Cancer and the Tropic of Capricorn tends to be warmer and experiences fewer variations in seasonal temperatures because it's constantly subject to the direct rays of the Sun, no matter which direction the Earth is tilted.

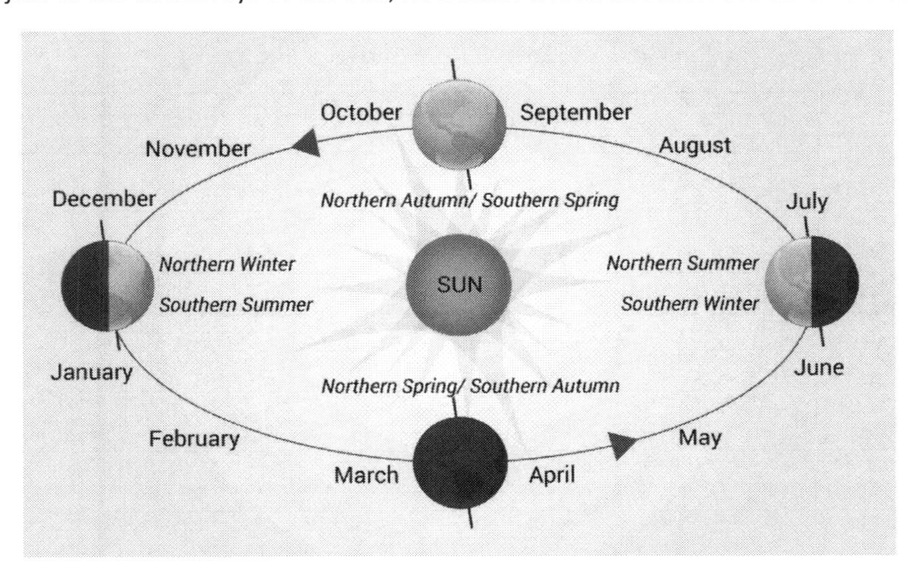

The area between the Tropic of Cancer and the Arctic Circle, which is at 66.5° North, and the Antarctic Circle, which is at 66.5° South, is where most of Earth's population resides and is called the **middle latitudes**. Here, the seasons are more pronounced, and milder temperatures generally prevail. When the Sun's direct rays are over the equator, it's known as an *equinox*, and day and night are almost equal throughout the world. Equinoxes occur twice a year: the fall, or autumnal equinox, occurs on September 22, while the spring equinox occurs on March 20.

Effects of Natural Phenomena

Natural phenomena can have a sizeable impact on climate and weather. Chemicals released from volcanic eruptions can fall back to Earth in acid rain. In addition, large amounts of carbon dioxide released into the atmosphere can warm the climate. Carbon dioxide creates the **greenhouse effect** by trapping solar energy from sunlight reflected off the Earth's surface within the atmosphere. The amount of solar radiation emitted from the Sun varies and has recently been discovered to be cyclical.

El Niño and La Niña

El Niño and **La Niña** are terms for severe weather anomalies associated with torrential rainfall in the Pacific coastal regions, mainly in North and South America. These events occur irregularly every few years, usually around December, and are caused by a band of warm ocean water that accumulates in the central Pacific Ocean around the equator. The warm water changes the wind patterns over the Pacific and stops cold water from rising toward the American coastlines. The rise in ocean temperature also leads to increased evaporation and rain. These events are split into two phases—a warm, beginning phase called El Niño and a cool end phase called La Niña.

Major Features of the Solar System

Structure of the Solar System

The **solar system** is an elliptical planetary system with a large sun in the center that provides gravitational pull on the planets.

Laws of Motion

Planetary motion is governed by three scientific laws called **Kepler's laws**:

1. The orbit of a planet is elliptical in shape, with the Sun as one focus.

2. An imaginary line joining the center of a planet and the center of the Sun sweeps out equal areas during equal intervals of time.

3. For all planets, the ratio of the square of the orbital period is the same as the cube of the average distance from the Sun.

The most relevant of these laws is the first. Planets move in elliptical paths because of gravity; when a planet is closer to the Sun, it moves faster because it has built up gravitational speed. As illustrated in the diagram below, the second law states that it takes planet 1 the same time to travel along the A1 segment as the A2 segment, even though the A2 segment is shorter.

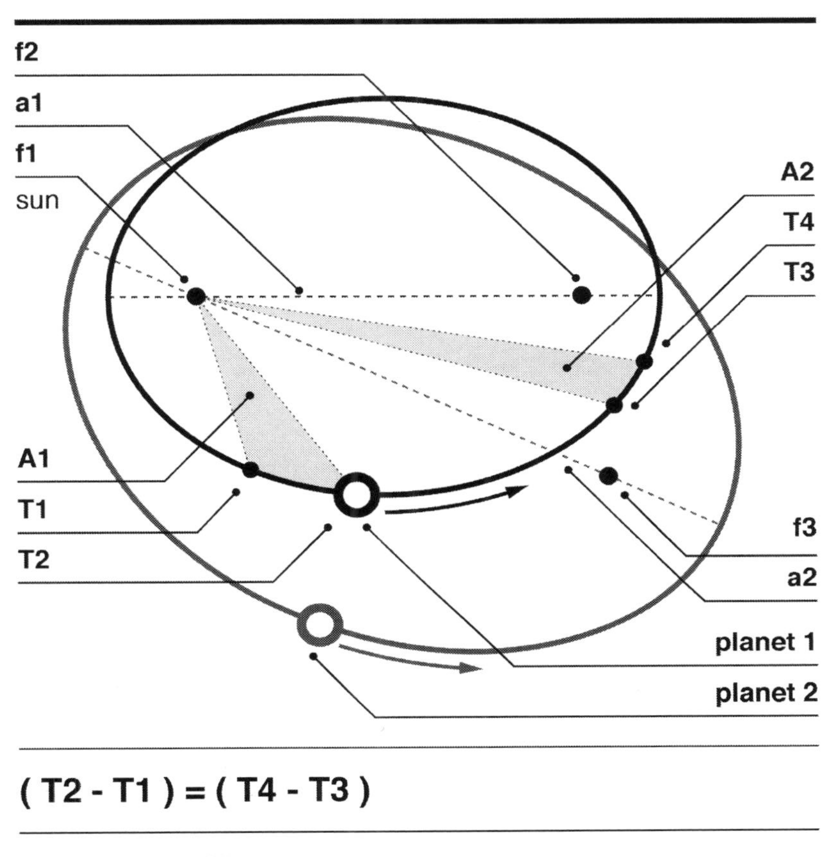

Characteristics of the Sun, Moon, and Planets

The Sun is comprised mainly of hydrogen and helium. Metals make up only about 2% of its total mass. The Sun is 1.3 million kilometers wide, weighs 1.989×10^{30} kilograms, and has temperatures of 5,800 Kelvin (9980 °F) on the surface and 15,600,000 Kelvin (28 million °F) at the core. The Sun's enormous size and gravity give it the ability to provide sunlight. The gravity of the Sun compresses hydrogen and helium atoms together through nuclear fusion and releases energy and light.

The Moon has a distinct core, mantle, and crust. It has elevations and craters created by impacts with large objects in the solar system. The Moon makes a complete orbit around the Earth every 27.3 days. It's relatively large compared to other moons in the Solar System, with a diameter one-quarter of the Earth and a mass 1/81 of the Earth.

The eight planets of the Solar System are divided into four inner (or **terrestrial**) planets and four outer (or **Jovian**) planets. In general, terrestrial planets are small, and Jovian planets are large and gaseous. The planets in the Solar System are listed below from nearest to farthest from the Sun:

- Mercury: the smallest planet in the Solar System; it only takes about 88 days to completely orbit the Sun
- Venus: around the same size, composition, and gravity as Earth and orbits the Sun every 225 days
- Earth: the only known planet with life
- Mars: called the Red Planet due to iron oxide on the surface; takes around 687 days to complete its orbit
- Jupiter: the largest planet in the system; made up of mainly hydrogen and helium
- Saturn: mainly composed of hydrogen and helium along with other trace elements; has 61 moons; has beautiful rings, which may be remnants of destroyed moons
- Uranus: the coldest planet in the system, with temperatures as low as -224.2 °Celsius (-371.56 °F)
- Neptune: the last and third-largest planet; also, the second-coldest planet

Asteroids, Meteoroids, Comets, and Dwarf/Minor Planets

Several other bodies travel through the universe. **Asteroids** are orbiting bodies composed of minerals and rock. They're also known as **minor planets**—a term given to any astronomical object in orbit around the Sun that doesn't resemble a planet or a comet. **Meteoroids** are mini-asteroids with no specific orbiting pattern. **Meteors** are meteoroids that have entered the Earth's atmosphere and started melting from contact with greenhouse gases. **Meteorites** are meteors that have landed on Earth. **Comets** are composed of dust and ice and look like a comma with a tail from the melting ice as they streak across the sky.

Theories of Origin of the Solar System

One theory of the origins of the Solar System is the **nebular hypothesis**, which posits that the Solar System was formed by clouds of extremely hot gas called a **nebula**. As the nebula gases cooled, they became smaller and started rotating. Rings of the nebula left behind during rotation eventually condensed into planets and their satellites. The remaining nebula formed the Sun.

Another theory of the Solar System's development is the **planetesimal hypothesis**. This theory proposes that planets formed from cosmic dust grains that collided and stuck together to form larger and larger bodies. The larger bodies attracted each other, growing into moon-sized protoplanets and eventually planets.

Interactions of the Earth-Moon-Sun System

The Earth's Rotation and Orbital Revolution Around the Sun

Besides revolving around the Sun, the Earth also spins like a top. It takes one day for the Earth to complete a full spin, or rotation. The same is true for other planets, except that their "days" may be shorter or longer. One Earth day is about 24 hours, while one Jupiter day is only about nine Earth hours, and a Venus day is about 241 Earth days. Night occurs in areas that face away from the Sun, so one side of the planet experiences daylight and the other experiences night. This phenomenon is the reason that the Earth is divided into time zones. The concept of time zones was created to provide people around the world with a uniform standard time, so the Sun would rise around 7:00 AM, regardless of location.

Effect on Seasons

The Earth's tilted axis creates the seasons. When Earth is tilted toward the Sun, the Northern Hemisphere experiences summer while the Southern Hemisphere has winter—and vice versa. As the Earth rotates, the distribution of direct sunlight slowly changes, explaining how the seasons gradually change.

Phases of the Moon

The Moon goes through two phases as it revolves around Earth: waxing and waning. Each phase lasts about two weeks:

- **Waxing**—the right side of the Moon is illuminated
    - **New moon** (dark): the Moon rises and sets with the Sun
    - **Crescent**: a tiny sliver of illumination on the right
    - **First quarter: the right half of the Moon is** illuminated
    - **Gibbous**: more than half of the Moon is illuminated
    - **Full moon**: the Moon rises at sunset and sets at sunrise
- **Waning**—the left side of the Moon is illuminated
    - **Gibbous**: more than half is illuminated, only here it is the left side that is illuminated
    - **Last quarter**: the left half of the Moon is illuminated
    - **Crescent**: a tiny sliver of illumination on the left
- **New moon** (dark)—the Moon rises and sets with the Sun

Effect on Tides

Although the Earth is much larger, the Moon still has a significant gravitational force that pulls on Earth's oceans. At its closest to Earth, the Moon's gravitation pull is greatest and creates high tide. The opposite is true when the Moon is farthest from the Earth: less pull creates low tide.

Solar and Lunar Eclipses

Eclipses occur when the Earth, the Sun, and the Moon are all in line. If the three bodies are perfectly aligned, a total eclipse occurs; otherwise, it's only a partial eclipse. A **solar eclipse** occurs when the Moon is between the Earth and the Sun, blocking sunlight from reaching the Earth. A **lunar eclipse** occurs when the Earth interferes with the Sun's light reflecting off the full moon. The Earth casts a shadow on the Moon, but the particles of the Earth's atmosphere refract the light, so some light reaches the Moon, causing it to look yellow, brown, or red.

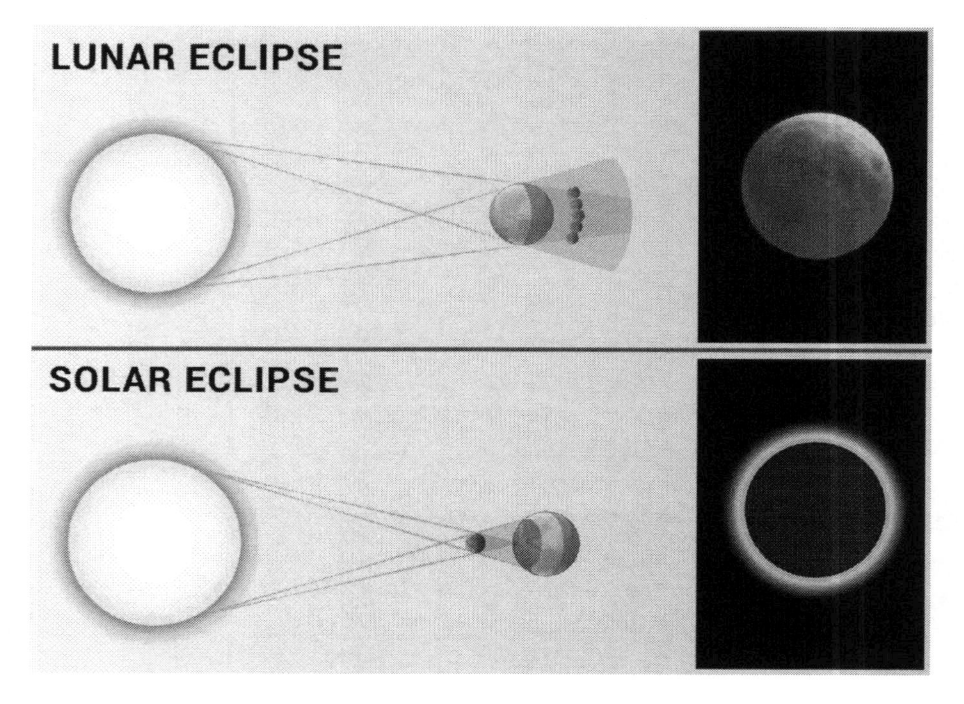

Time Zones

Longitudinal, or vertical, lines determine how far east or west different regions are from each other. These lines, also known as **meridians**, are the basis for time zones, which allocate different times to regions depending on their position eastward and westward of the prime meridian.

Effect of Solar Wind on the Earth

Solar winds are streams of charged particles emitted by the Sun, consisting of mostly electrons, protons, and alpha particles. The Earth is largely protected from solar winds by its magnetic field. However, the winds can still be observed, as they create phenomena like the beautiful Northern Lights (or **Aurora Borealis**).

Major Features of the Universse

Galaxies

Galaxies are clusters of stars, rocks, ice, and space dust. Like everything else in space, the exact number of galaxies is unknown, but there could be as many as a hundred billion. There are three types of galaxies: spiral, elliptical, and irregular. Most galaxies are **spiral galaxies**; they have a large, central galactic bulge made up of a cluster of older stars. They look like a disk with spinning arms. **Elliptical galaxies** are groups of stars with no pattern of rotation. They can be spherical or extremely elongated, and they don't have arms. **Irregular galaxies** vary significantly in size and shape.

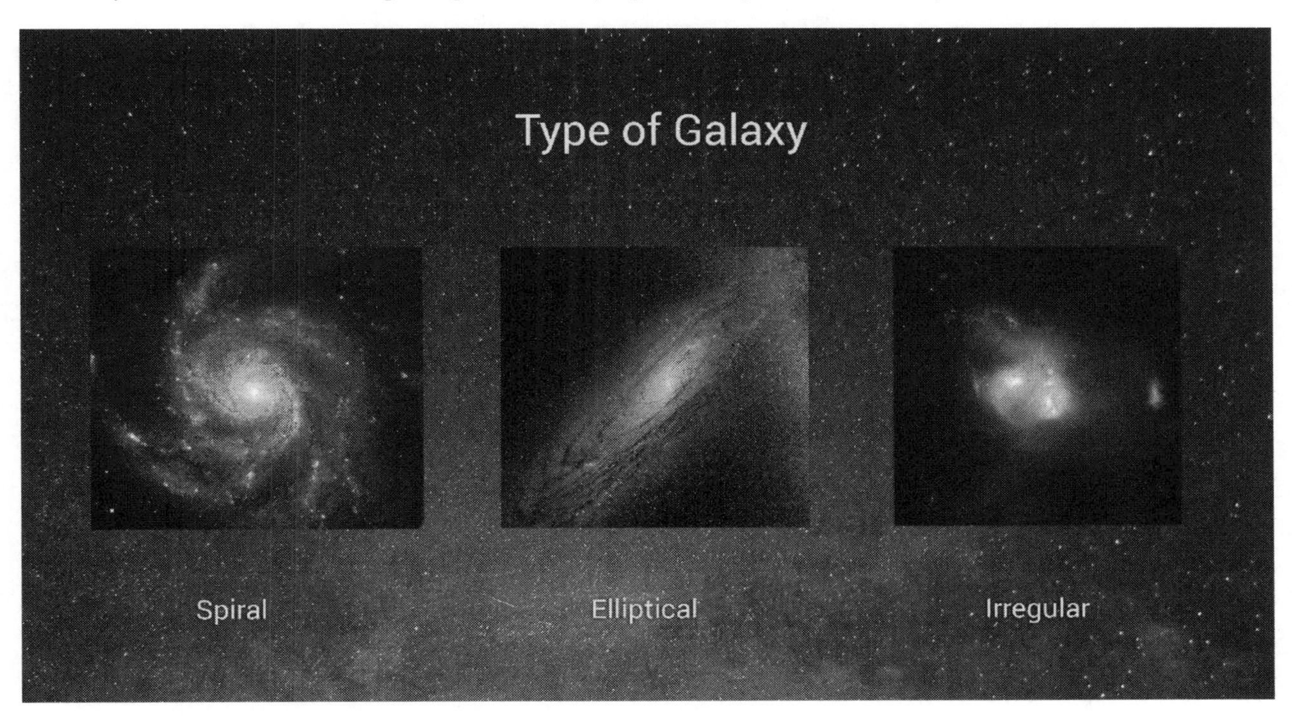

To say that galaxies are large is an understatement. Most galaxies are 1,000 to 100,000 parsecs in diameter, with one *parsec* equal to about 19 trillion miles. The Milky Way is the galaxy that contains Earth's Solar System. It's one of the smaller galaxies that has been studied. The diameter of the Milky Way is estimated to be between 31,000 to 55,000 parsecs.

Characteristics of Stars and Their Life Cycles

Life Cycle of Stars

All stars are formed from **nebulae**. Depending on their mass, stars take different pathways during their life. Low- and medium-mass stars start as nebulae and then become red giants and white dwarfs. High-mass stars become red supergiants, supernovas, and then either neutron stars or black holes. Official stars are born as red dwarves because they have plentiful amounts of gas—mainly hydrogen—to undergo nuclear fusion. Red dwarves mature into white dwarves before expending their hydrogen fuel source. When the fuel is spent, it creates a burst of energy that expands the star into a red giant. Red giants eventually condense to form white dwarves, which is the final stage of a star's life.

Stars that undergo nuclear fusion and energy expenditure extremely quickly can burst in violent explosions called **supernovas**. These bursts can release as much energy in a few seconds as the Sun can release in its entire lifetime. The particles from the explosion then condense into the smallest type of star—a neutron star—and eventually form a **black hole**, which has such a high amount of gravity that not even light energy can escape. The Sun is currently a red dwarf, early in its life cycle.

Color, Temperature, Apparent Brightness, Absolute Brightness, and Luminosity

The color of a star depends on its surface temperature. Stars with cooler surfaces emit red light, while the hottest stars give off blue light. Stars with temperatures between these extremes, such as the Sun, emit white light. The **apparent brightness** of a star is a measure of how bright a star appears to an observer on the Earth. The **absolute brightness** is a measure of the intrinsic brightness of a star and is measured at a distance of exactly 10 parsecs away. The **luminosity** of a star is the amount of light emitted from its surface.

Hertzsprung-Russell Diagrams

Hertzsprung-Russell diagrams are scatterplots that show the relationship of a star's brightness and temperature, or color. The general layout shows stars of greater luminosity toward the top of the diagram. Stars with higher surface temperatures appear toward the left side of the diagram. The diagonal area from the top-left of the diagram to the bottom-right is called the **main sequence**. Stars may or may not follow the main sequence during their life.

Dark Matter

Dark matter is an unidentified type of matter that comprises approximately 27% of the mass and energy in the observable universe. As the name suggests, dark matter is so dense and small that it doesn't emit or interact with electromagnetic radiation, such as light, making it electromagnetically invisible. Although dark matter has never been directly observed, its existence and properties can be inferred from its gravitational effects on visible objects as well as the cosmic microwave background. Patterns of movement have been observed in visible objects that would only be possible if dark matter exerted a gravitational pull.

Theory About the Origin of the Universe

The **Big Bang theory** is a proposed cosmological model for the origin of the universe. It theorizes that the universe expanded from a high-density and high-temperature state. The theory offers comprehensive explanations for a wide range of astronomical phenomena, such as the cosmic microwave background and Hubble's Law. From detailed measurements of the expansion rate of the universe, Big Bang theorists estimate that the Big Bang occurred approximately 13.8 billion years ago, which is considered the age of the universe. The theory states that after the initial expansion, the universe cooled enough for subatomic particles and atoms to form and aggregate into giant clouds. These clouds coalesced through gravity and formed the stars and galaxies. If this theory holds true, it's predicted that the universe will reach a point where it will stop expanding and start to pull back toward the center due to gravity.

Contributions of Space Exploration and Technology to our Understanding of the Universe

Remote Sensing Devices

Scientists and astronomers use satellites and other technology to explore the universe because humans cannot yet safely travel far into space. The *telescope* allows observation beyond what the naked eye can see. With information from the Planck satellite, astronomers were able to determine that the observable universe is actually smaller than earlier believed; they estimated that the universe is smaller by 320 million light years, giving it a total radius of 45.34 billion light years. **Astronomical spectroscopy** uses spectroscopy techniques to measure visible light and radiation from stars and other hot celestial objects. This information allows scientists to determine the physical properties of stars that would otherwise be impossible to measure, including chemical composition, temperature, density, and mass.

Search for Water and Life on Other Planets

Even the smallest microorganisms on Earth cannot live without liquid water. Water is essential to life, which is why scientists believe the search for water is the best way to find life on other planets. Water can be found throughout the universe in the form of ice. For example, some of Saturn's rings are composed of ice, and comets streaking through the sky release ice particles through their tails. Until recently, it was believed that the Earth was the only place with liquid water. Through the use of satellites and telescopes, astronomers have discovered that two of Jupiter's moons—Europa and Callisto—may contain liquid water below their surfaces.

Physics

Nature of Motion

People have been studying the movement of objects since ancient times, sometimes prompted by curiosity, and sometimes by necessity. On Earth, items move according to specific guidelines and have motion that is fairly predictable. The measurement of an object's movement or change in position (x), over a change in time (t) is an object's **speed.** The average speed of an object is defined as the distance that the object travels divided by how long it takes the object to travel that distance. When the direction is included with the speed, it is referred to as the **velocity**. A "change in position" refers to the difference in location of an object's starting point and an object's ending point. In science, this change is represented by the Greek letter Delta, Δ.

$$velocity\ (v) = \frac{\Delta x}{\Delta t}$$

Distance is measured in meters, and time is measured in seconds. The standard scientific units for speed and velocity are meters/second (m/s), but units of miles/hour (mph) are also commonly used in America.

$$\frac{meters}{second} = \frac{m}{s}$$

Average velocity is calculated by averaging the beginning or initial velocity and the ending or final velocity of an object.

If a measurement includes its direction, it is called a **vector quantity**; otherwise, the measurement is called a **scalar quantity** and has only a numeric value without a particular direction. For example, speed is a scalar quantity while velocity is a vector quantity.

Acceleration

While an object's speed measures how fast the object's position will change in a certain amount of time, an object's **acceleration** measures how fast the object's speed will change in a certain amount of time.

Acceleration can be thought of as the change in velocity or speed (Δv) divided by the change in the time (Δt).

$$acceleration\ (a) = \frac{\Delta v}{\Delta t}$$

Velocity is measured in meters/second and time is measured in seconds, so the standard unit for acceleration is meters / second2 (m/s2).

$$\frac{meters/_{second}}{second} = \frac{meters}{second^2} = \frac{m}{s^2}$$

Acceleration is expressed by using both magnitude and direction, so it is a vector quantity like velocity. Acceleration is present when an object is slowing down, speeding up, or changing direction, since these represent instances where velocity is changing. This means that forces like friction and gravity accelerate objects and increase or decrease their velocities over time.

Projectile Motion

Projectile motion describes the way in which a projectile will move when the only force acting upon it is gravity. Since the force of Earth's gravity is nearly constant at sea level, its magnitude is approximated as a rate of 9.8 m/s2. For projectile motion problems, the projectile is assumed to travel a curved path to the ground and ignore air resistance, wind speed, and other such complications.

Projectile motion has two components: horizontal and vertical. Without air resistance, there is no horizontal acceleration, so the horizontal velocity won't change. However, the vertical velocity will change because of gravity's acceleration on the projectile. A sample parabolic curve of projectile motion is shown below.

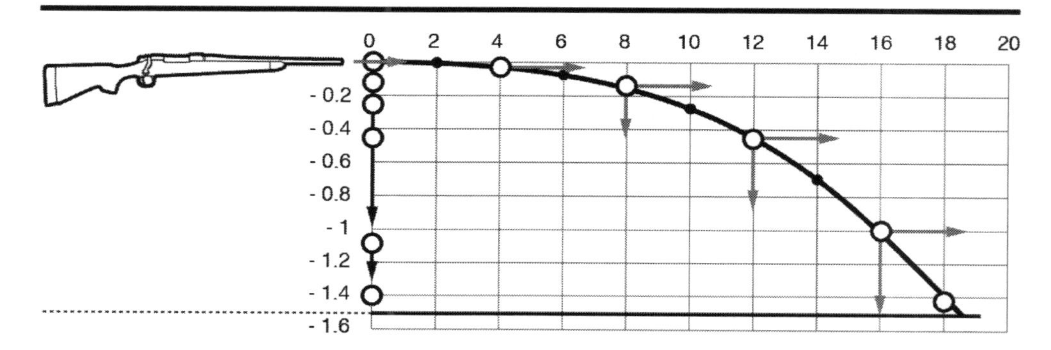

Horizontal distance (d_x) is defined as the relationship between velocity (v_x) and time (t), where x is the movement along the horizontal plane (x-axis).

$$d_x = v_x t$$

Because of the force of gravity, the vertical velocity and position is continuously changing and thus more complicated to calculate. The following equations are different expressions of vertical motion:

$$v_f^{\,2} = v_i^{\,2} + 2ad$$

$$d = \frac{1}{2}at^2 + v_i t$$

$$v_f = v_i + at$$

These equations use v_f as the final velocity, v_i as the initial velocity, a as the acceleration, d as the horizontal distance traveled, and t as the time to describe functions of motion.

Newton's Laws of Motion

Sir Isaac Newton spent a great deal of time studying objects, forces, and how an object's motion responds to forces. Newton made great advancements by using mathematics to describe the motion of objects and to predict future motions of objects by applying his mathematical models to situations. Through his extensive research, Newton is credited for summarizing the basic laws of motion for objects here on Earth. These laws are as follows:

1. The law of inertia: An object in motion will remain in motion, unless acted upon by an outside force. An object at rest remains at rest, unless acted upon by an outside force. Simply put, **inertia** is the natural tendency of an object to continue along with what it is already doing; an outside force would have to act upon the object to make it change its course. This includes an object that is sitting still.

2. F = ma: The **force (F)** on an object is equal to the mass (m) multiplied by the acceleration (a) on that object. **Mass (m)** refers to the amount of a substance while acceleration (a) refers to a change in velocity over time.

    - In the case of a projectile falling to Earth's surface, the acceleration is due to gravity.
    - Multiplying an object's mass by its gravitational acceleration gives the special force that is called the object's **weight (W).** Note that weight is a force and is a vector quantity while mass is in kilograms and is a scalar quantity, as it has no acceleration.
    - Forces are typically measured in Newtons (N), which are a derived SI unit equal to:

    $$1\,\frac{kg{\cdot}m}{s^2}$$

3. For every action, there is an equal and opposite reaction. This means that, if a book drops onto a desk, the book will exert a force on the desk due to hitting it, but the desk will also exert an equal force on the book in the opposite direction. This is what sometimes causes fallen objects to bounce once or twice after hitting the ground.

A clear understanding of force is crucial to using Newton's laws of motion. Forces are anything acting upon an object either in motion or at rest; this includes friction and gravity. These forces can push or pull on a mass, and they have a magnitude and a direction. Forces are represented by a **vector,** which is the arrow lined up along the direction of the force with its tip at the point of application. The magnitude of the force is represented by the length of the vector: Large forces have long vector lengths.

Friction

Friction is a resistance to movement that always imparts a negative acceleration to an object. Because it accelerates an object in the opposite direction than it wants to go, friction will cause moving objects to slow down and eventually stop. Frictional force depends on several factors including the texture of the two surfaces and the amount of contact force pushing the surfaces together. There are four types of friction: static, sliding, rolling, and fluid friction. **Static friction** occurs between stationary objects, while **sliding friction** occurs when solid objects slide over each other. **Rolling friction** happens when a solid object rolls over another solid object, and **fluid friction** is a friction caused by an object moving through a fluid or through fluid layers. These 'fluids' can be either a gas or a liquid, so this also includes air resistance.

Rotation

Rotating and spinning objects have a special type of movement. A spinning top can move across a table, demonstrating linear motion, but it can also spin in place, demonstrating angular motion. Just as a car moves from place to place by changing its location, its velocity, and its acceleration, so too can a rotating object change its orientation, its angular velocity, and its angular acceleration.

For linear movement, the first thing that is described is the object's displacement. The **displacement** of the object is how far it moves from its starting location; this linear change in location is usually represented by the symbol Δx. However, for the angular movement of a simple solid like a sphere, the distance that its surface travels isn't a good measure of its angular movement, since a large sphere would have to rotate a much smaller degree to move the same distance as a small sphere would. This angular distance is referred to as S, or the arc length.

A better measure of angular movement is the **angular displacement** θ, which is defined as the angle through which the object will rotate. Because there are a standard 360° in all circles, angular displacement can be used to compare angular movement between objects of different sizes, making it a much more versatile tool than the arc length. However, although a circle can be split up into 360 degrees, it can also be visualized as being split up into 2π **radians,** where radians is a unit similar to degrees that describes angles. The 2π is usually used in physics because a circle's circumference has a value of 2π times its radius, and radians is an easier unit to use than is degrees.

The angular displacement can be found by dividing the arc length that an object rotates through by the radius of the object's rotation. For example, if an object completes 1 full rotation, it has rotated through 360°. It has also traveled an arc length equal to its circumference, or $2\pi r$. Plugging this arc length S into the equation below, it can be seen that the angular displacement is 2π, which is correct since it is equal to the 360° as discussed previously.

$$angular\ displacement\ (\theta) = \frac{S}{r}$$

The angular speed or **angular velocity, w,** is the measure of how quickly the object is rotating, and w is defined as the angular displacement that is accomplished in a certain amount of time, as shown below.

$$angular\ speed\ (\omega) = \frac{\Delta\theta}{\Delta t}$$

Similar to linear velocity, angular velocity may accelerate due to external forces, and this angular acceleration is given by the variable α. Its equation, the change in angular velocity over a change in time, is given below.

$$angular\ acceleration\ (\alpha) = \frac{\Delta\omega}{\Delta t}$$

When objects are exhibiting circular motion, they also demonstrate the **conservation of angular momentum,** meaning that the angular momentum of a system is always constant, regardless of the placement of the mass. **Rotational inertia** can be affected by how far the mass of the object is placed with respect to the center of rotation (**axis of rotation**). The larger the distance between the mass and the center of rotation, the slower the rotational velocity. Conversely, if the mass is closer to the center of rotation, the rotational velocity increases. A change in one affects the other, thus conserving the angular momentum. This holds true if no external forces act upon the system.

Circular Motion

Circular motion is similar in many ways to linear (straight line) motion; however, there are a few additional points to note. In uniform circular motion, a spinning object is always linearly accelerating because it is always changing direction. The force causing this constant acceleration on or around an axis is called the **centripetal force,** and it is often associated with centripetal acceleration. Centripetal force always pulls toward the axis of rotation; this means that the force will always pull towards the center of the rotation circle. The relationship between the velocity (v) of an object and the radius (r) of the circle is centripetal acceleration, and the equation is as follows:

$$centripetal\ acceleration\ (a_c) = \frac{v^2}{r}$$

According to Newton's law, force is the combination of two factors, mass and acceleration. This is demonstrated by centripetal force. Centripetal force is shown mathematically by using the mass of an object (m), the velocity (v), and the radius (r).

$$centripetal\ force\ (F_c) = \frac{mv^2}{r}$$

Kinetic and Potential Energy

The two primary forms of energy are kinetic energy and potential energy. **Kinetic energy**, or KE, involves the energy of motion, and is easily found for Newtonian physics by an object's mass in kilograms and velocity in meters per second. Kinetic energy can be calculated using the following equation:

$$KE = \frac{1}{2}mv^2$$

Potential energy represents the energy possessed by an object by virtue of its position. In the classical example, an object's gravitational potential energy can be found as a simple function of its height or by what distance it drops. Potential energy, or PE, may be calculated using the following equation:

$$PE = mgh$$

Both kinetic energy and potential energy are scalar quantities measured in **Joules**. One Joule is the amount of energy that can push an object with 1 Newton of force for 1 meter, so it is also referred to as a Newton-meter. As mentioned previously, the **Law of Conservation of Energy** states that energy can neither be created nor destroyed. Therefore, potential and kinetic energy can be transformed into one another, depending on an object's speed and position.

Linear Momentum and Impulse

Motion creates something called **momentum.** This is a calculation of an object's mass multiplied by its velocity. Momentum can be described as the amount an object wants to continue moving along its current course. Momentum in a straight line is called linear momentum. Just as energy can be transferred and conserved, so too can momentum.

Changing the expression of Newton's second law of motion yields a new expression.

$$Force(F) = ma = m \times \frac{\Delta v}{\Delta t}$$

If both sides of the expression are multiplied by the change in time, the law produces the impulse equation.

$$F\Delta t = m\Delta v$$

This equation shows that the amount of force during a length of time creates an **impulse.** This means that if a force acts on an object during a given amount of time, it will have a determined impulse. However, if the same change in velocity happens over a longer amount of time, the required force is much smaller, due to the conservation of momentum.

$$p = mv$$

Linear momentum, p, is found by multiplying the mass of an object by its velocity. Since momentum, like mass and energy, is conserved, Newton's 2nd law can be restated for multiple objects. In this form, it can be used to understand the energy of objects that have interacted, since the conservation of momentum implies that the momentum before and after an interaction must be the same. This is best demonstrated in the case of elastic collision, where an object of mass m_1 with velocity v_1 collides with an object of mass m_2 with velocity v_2 and both object end with velocities v_1' and v_2', respectively.

$$m_1 v_1' + m_2 v_2' = m_1 v_1 + m_2 v_2$$

Universal Gravitation

Newton's **law of universal gravitation** addresses the universality of gravity. Gravity acts as a force at a distance and causes all bodies in the universe to attract each other.

The **force of gravity (F_g)** is proportional to the masses of two objects (m_1 and m_2) and inversely proportional to the square of the distance (r^2) between them. (G is the proportionality constant). This is shown in the following equation:

$$F_g = G\frac{m_1 m_2}{r^2}$$

All objects falling within the Earth's atmosphere are all affected by the force of gravity, so their rates of acceleration will be equal to 9.8 m/s^2, or gravity. Therefore, if two objects are dropped from the same height at the same time, they should hit the ground at the same time. This is irrespective of mass, since the previous kinetics equations don't include mass, so a bowling ball and a feather would theoretically fall at identical rates. Unfortunately, in the Earth's atmosphere, air resistance would slow the feather much more than the bowling ball, so the bowling ball would fall faster, but this effect can be minimized in a vacuum. In other words, without air resistance or other external forces acting on the objects, gravity will affect every object on the Earth with the same rate of acceleration.

Waves and Sound

Waves are periodic disturbances in a gas, liquid, or solid that are created as energy is transmitted. Each part of a wave has a different name and is used in different calculations. The four parts of a wave are the crest, the trough, the amplitude, and the wavelength. The **crest** is the highest point, while the **trough** is the lowest. The **amplitude** is the distance between a peak and the average of the wave; it is also the distance between a trough and the average of the wave, but an amplitude is always positive, since it is an absolute value. Finally, the distance between one wave and the exact same place on the next wave is the **wavelength.**

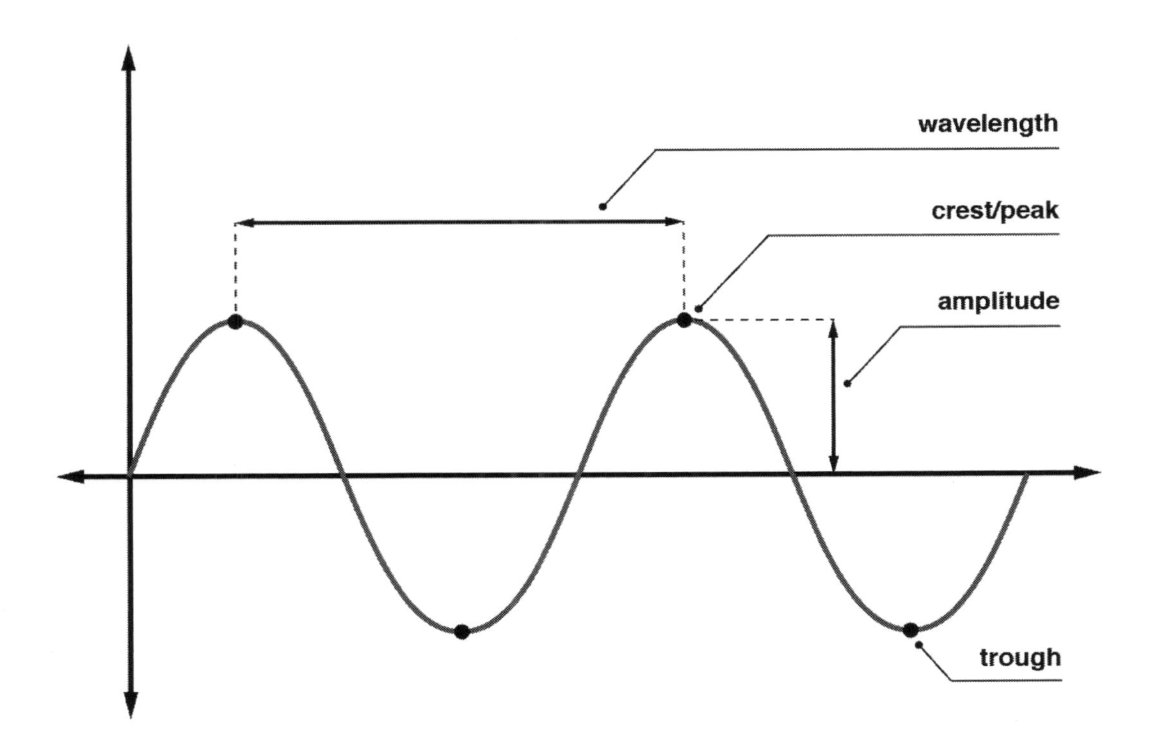

With amplitude and wavelength, it is possible to describe any wave, but an important question still remains unanswered: How fast is the wave traveling? A wave's speed can be shown as either its period or its frequency. A wave's period, T, is how long it takes for the wave to travel one wavelength, while a wave's frequency, f, is how many wavelengths are traveled in one second. These are inversely related, so they are reciprocals of each other, as shown below.

$$f = \frac{1}{T} \text{ and } T = \frac{1}{f}$$

The largest categories of waves are electromagnetic waves and mechanical waves. **Electromagnetic waves can transmit energy through a vacuum and do not need a medium to travel through, examples of which are light, radio, microwaves, gamma rays, and other forms of electromagnetism. Mechanical waves** can only transmit energy through another form of matter called a medium. The particles of the medium are shifted as the wave moves through the medium and can be anything from solids to liquids to gasses. Examples of mechanical waves include auditory sounds heard by human ears in the air as well as percussive shocks like earthquakes.

There are two different forms of waves: transverse and longitudinal waves. **Transverse** waves are waves in which particles of the medium move in a direction perpendicular to the direction waves move, as in most electromagnetic waves. **Compression, or longitudinal, waves are waves in which the particles of the medium move in a direction parallel to the direction the waves move, as in most mechanical waves. A good example of a longitudinal wave is sound.** Waves travel within a medium at a speed that is determined by the wavelength (λ) and frequency (f) of the wave.

$$v = f\lambda$$

There is a proportional relationship between the amplitude of a wave and the potential energy in the wave. This means the taller the wave, the more stored energy it is transmitting.

Light

Light is an electromagnetic wave that is created by electric and magnetic interactions. Like other electromagnetic waves, light does not need a medium to travel. Light energy can be absorbed and changed into heat, reflected, or even transmitted.

A wave is reflected when it collides with a surface and bounces off, unharmed. The **law of reflection** shows that an **incident ray** (the wave that hits the surface) will bounce off the surface and become a reflected ray (the wave that leaves the surface). Because the wave doesn't lose any energy, the angle at which it hits the surface will be identical to the angle at which it leaves the surface, so the interaction produces identical waves. This is the definition of **reflection**.

On the other hand, a wave is refracted when a wave collides with a surface and bends, such as when the medium it is traveling through changes. Examples of **refractio**n include the bending of light as it passes through a prism and sunlight passing through raindrops to create a rainbow.

To understand refraction, a 'normal' line is drawn perpendicular to the surface that the wave will hit. The angle created between the normal line and the incident ray is the angle of incidence, while the angle of reflection is the opposite and equal angle formed between the normal line and the reflected ray. The refraction of light depends on the varying speeds and densities of different media. As light passes through different media, the wave will bend toward or away from the normal line, depending on the material it is transitioning to.

Snell's law describes this behavior mathematically in the equation that follows:

$$n_1 \sin \theta_1 = n_2 \sin \theta_2$$

In this equation, n is the index of refraction and θ is the angle of refraction. The **index of refraction** determines the amount of light that bends or refracts when it encounters a medium. Materials that interact with and change the wave more will have higher indices of refraction, and any material's index of refraction can be found by the following equation.

$$n = \frac{c}{v_s}$$

Because the speed of light in a vacuum, c, is constant, it can be used to show how much a material will interact with a wave. The **refractive index, *n*,** shows how much the speed of light is changed when it travels through the material at a new velocity of v_s.

Optics

Spherical mirrors change the way light reflects. Lenses and curved mirrors are made to focus light in certain ways. There are several terms used to describe and define mirrors and lenses. The **principal axis** is a reference line that usually passes through the center of the curve of the mirror or lens. The **vertex, V,** is the point where the mirror is crossed by the principal axis. The sphere which contains the curve of the mirror has a center called the **center of curvature (C)**. The **focal point (F)** is halfway between the center of curvature and the vertex. The **focal length (f)** is measured by how far the focal point is from the mirror. The following graphic shows a visual representation of these terms in a concave mirror.

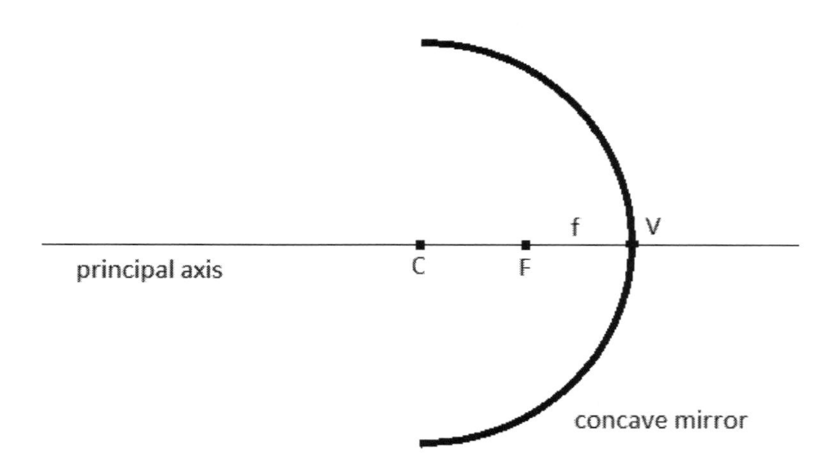

The focal lengths of concave mirrors are positive. Concave mirrors can produce images of various shapes and sizes as well as of different orientations based on the focal length of the mirror and the placement of the object. Convex mirrors differ from concave mirrors in that they have negative focal lengths. The other main difference is that convex mirrors always form images that are reduced in size and upright.

Lenses are pieces of transparent material molded to refract light rays to create an image. There are two types of lenses. Convex lenses, also called **converging lenses**, have positive focal lengths. Concave lenses, or **diverging lenses**, conversely have negative focal lengths. Images in convex lenses can look different depending on the focal length of the lens and where the object is located. Convex lenses only create images that are upright and smaller in size than the object.

Take a look at this image:

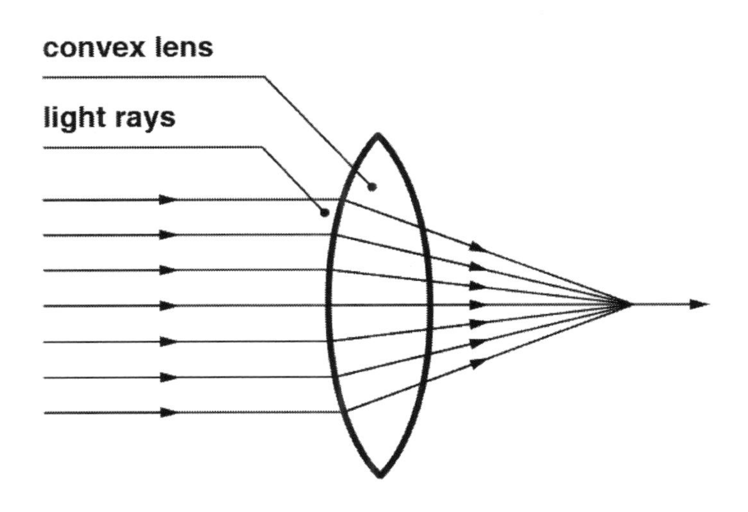

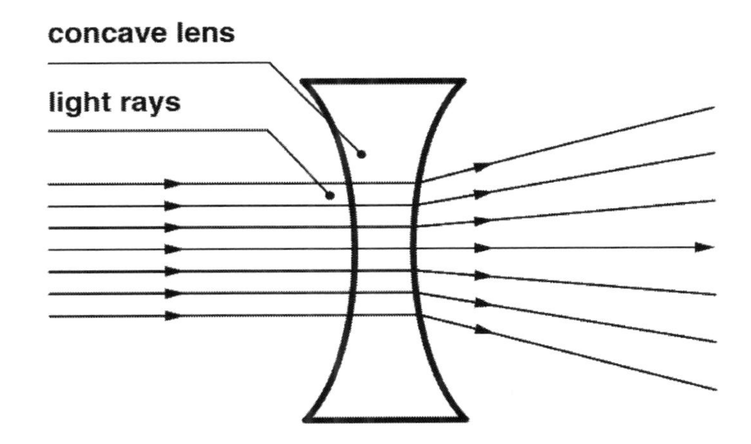

Atomic Structure

All known matter is made of atoms. Atoms have a nucleus composed of **protons** (with a positive charge) and **neutrons** (with no net charge), which is surrounded by a cloud of **electrons** (with negative charges). Since protons have a positive charge and neutrons have a neutral charge, an atom's nucleus typically has an overall positive electrical charge, but because electrons orbiting the nucleus have an overall negative charge, an atom with equal numbers of protons and electrons is considered stable.

The properties of an atom vary with the number of electrons and their arrangement in shells around the nucleus. Electrons are organized into distributions of subshells called **electron configurations**. Subshells fill from the inside (closest to the nucleus) to the outside and are lettered starting with "k." The strength of the bond between the nucleus and the electron is called the **binding energy**. The binding energy is stronger for those shells located closer to the nucleus.

The number of electrons in each shell can be determined using the following equation using n for the shell number:

$$2n^2$$

Nature of Electricity

The nature of electricity is based on the atoms of a given material, as objects develop electric charges when their atoms gain or lose electrons. The electrons that are in the furthest shell from an atom's nucleus are the most likely to interact, so they are specially termed valence electrons. If the electrons are tightly bound to their atoms, the material is an **insulator** because, like rubber, wood, and glass, the material prevents electrons from flowing freely. However, if the electrons are loosely bound to the atoms, the material is a **conductor** because, like iron and other metals, the electrons can flow freely throughout the material.

An atom can gain a charge by having greater or fewer electrons than protons, and any atom with a charge is referred to as an **ion**. If an ion has more electrons than protons, it has a negative charge and is an **anion**. If an atom has fewer electrons than protons, it has a positive charge and is a **cation**. It is important to note that opposite charges attract each other, while like charges repel each other. Interestingly, the repulsive force between two electrons is equivalent to the attractive force between an electron and a proton (only in different directions). Coulomb described these repellant and attractive forces between two objects in his namesake law. This electrical **Coulomb force, F,** is determined by multiplying the charges of the two objects, q_1 and q_2, by a constant, k, and dividing by the distance they are from one another squared, r^2.

$$F = \frac{kq_1q_2}{r^2}$$

If the force is negative, then the objects attract each other because q_1 and q_2 are of opposite charges. If the force is positive, then the objects repel each other because q_1 and q_2 are of the same charge.

An electric charge naturally produces an electric field around itself. Electric fields can vary in strength and magnitude depending on the type of charge (positive or negative) that generates the field. The nature of electric fields can be tested using a test charge. Mathematically, the magnitude of an electric field (E) can be found using the following equation:

$$E = \frac{F}{q_o}$$

Where F is the force a test charge would undergo, and q_o is the magnitude of the test charge. Electric fields are vector quantities; this means they have both a magnitude and a direction. In the case of electric fields, it is the convention to define the direction of the field vector as the way a positive test charge would move when positioned in the electric field, towards a negative charge and away from other positive charges.

An **electric circuit** is usually comprised of circuit elements joined by a wire or other object that allows an electric charge to move along the path without interruption—this moving electric charge is called an electric **current.** However, constant electric currents may only exist in a complete circuit if there is a voltage difference in the circuit. **Voltage** is literally the distance that a circuit's electrical force could move one electron, but voltage can be visualized as how much energy a certain part of the circuit has available to push around electrons. Electrons will flow from regions of higher voltage to regions of lower voltage, so it is the difference in voltages between two parts of a circuit that makes a current actually flow.

In other words, voltage difference is the difference in potential energy between two places measured in **volts (V),** while a circuit is a closed path through which electrons can flow. Because every atom has positive charges that pull on electrons and resist their flow, most real circuits have a **resistance** level, measured in **ohms**, which is described as the opposition to the flow of electric charge. The amount of current that can flow through a circuit depends on the voltage difference and how well the wire resists the flow of electricity. **Ohm's law** gives us the relationship between voltage (V), current in amperes (I), and resistance in ohms (R).

$$V = I \times R$$

Series Circuits (A) and Parallel Circuits (B)

Practical circuits have numerous loads which can be hooked up "in series" (A) or "in parallel" (B), as shown below.

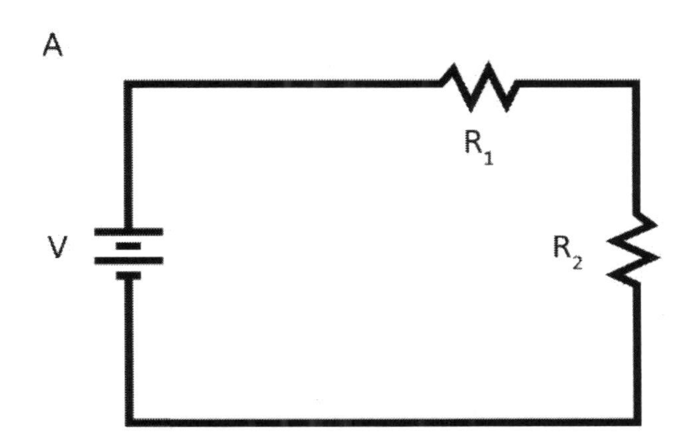

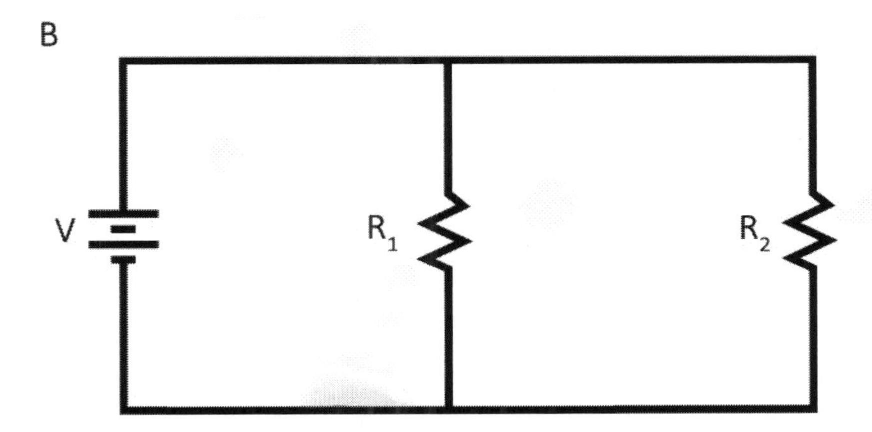

To determine the total voltage requirement for circuits with multiple component loads (in series or in parallel), it is necessary to find the equivalent resistance of the circuits.

Series circuits put resistors in a row or series so that current must flow from one to the other, while **parallel circuits** run resistors in parallel sections so that current can flow through one or the other. In a series circuit like the one in *A*, the voltage drops across each resistor, but the current is the same in all of them. The current must be the same across each resistor, as according to Ampere's law, the electrons in the current must continue flowing throughout the wire and not build up or disappear. In other words, the electrons going into the resistor must all go out so that the "flow in equals flow out."

The current through each resistor is the same, and the total voltage (V) equals the drop across R_1 plus the drop across R_2. The equivalent resistance is determined by solving Ohm's law for voltage:

$$V = V_1 + V_2 = IR_1 + IR_2 = I(R_1 + R_2) = IR_{eq}$$

Thus, in a series circuit, the equivalent resistance is equal to the sum of the component resistances, and this relation holds for any number of resistors in a series.

In a parallel circuit like the one in *(B)*, the voltage is the same across each resistor because each is attached directly to the power source and the ground. The electric current is divided between the loads depending on their resistances, since it can flow through either of them, but not both. If the resistance is the same in both loads, then the same amount of current passes through each one. If the resistance is different in each load, then more current passes through the load with the lower resistance, since energy takes the path of least resistance.

The **equivalent resistance (R_{eq})** of the parallel circuit is determined by solving Ohm's law for the current through each resistor, setting it equal to the total current (R_t), and remembering that the voltages are all the same:

$$I_t = \frac{V}{R_{eq}} = \frac{V}{R_1} + \frac{V}{R_2} \quad \textit{or} \quad \frac{1}{R_{eq}} = \frac{1}{R_1} + \frac{1}{R_2} \quad \textit{so} \quad R_{eq} = \frac{1}{\frac{1}{R_1} + \frac{1}{R_2}}$$

In a parallel circuit, the equivalent resistance is equal to one over the sum of the reciprocals of the component resistances.

Magnetism and Electricity

Magnetism can occur naturally in certain types of materials like iron, nickel, and cobalt. If two straight rods are made from iron, they will usually have a naturally negative end (pole) and a positive end (pole). These charged poles react just like any charged item: opposite charges attract and like charges repel. They will attract each other when set up positive to negative, but if one rod is turned around, the two rods will repel each other due to the alignment of negative to negative and positive to positive.

Magnetic fields can also be created and amplified by using an electric current. The force of attraction between two magnetic fields is measured in **Teslas**. The relationship between magnetic forces and electrical forces can be explored by sending an electric current through a stretch of wire, which creates an electromagnetic force around the wire from the charge of the current, as long as the flow of electricity is sustained. This magnetic force can also attract and repel other items with magnetic properties. Depending upon the strength of the current in the wire, a smaller or larger magnetic force can be generated around this wire. As soon as the current is cut off, the magnetic force also stops. When a magnetic field produces an electric current, this is called an **electromagnetic induction**.

Types of Passages and Tips

The ACT science exam tests analytical skills associated with interpreting scientific principles of the natural world. The content may be from biology, the earth sciences, chemistry, or physics. The test presents scientific concepts in the form of passages of three different formats. A total of 40 multiple-choice questions related to the passage are to be completed in 35 minutes.

Test-takers are often surprised to discover that detailed knowledge of science content isn't required to achieve success on the ACT science test. In fact, the test is designed to determine how well an individual is able to analyze, compare, and generalize information. For example, the test doesn't seek to determine if one possesses an extensive knowledge of elephant biology. Instead, the exam might ask one to compare two scientists' differing opinions about elephant communication. In this case, the test taker is *not* required to possess any knowledge about elephants or the manner in which they communicate. All of the information needed to answer the questions will be presented in the passage. The test taker must read the passage, compare the differing hypotheses, analyze the graphical information, and generate conclusions.

Three Types of Passages

Data Representation

Scientific data will be presented in tables, graphs, diagrams, or models. These passages account for 30% to 40% of the ACT Science Test. These passages typically contain 5 or 6 questions designed to test one's ability to interpret scientific data represented in a graphical format instead of written text. These questions do *not* necessarily examine scientific content knowledge (e.g., the equation for photosynthesis); rather, they test students' ability to interpret raw data represented in a table or graph. Therefore, it's possible to do well on this portion of the exam without a detailed understanding of the scientific topic at hand. Questions may ask for factual information, identification of data trends, or graph calculations. For example:

- Based on the attached graph, how did Study 1 differ from Study 2?
- What is the nature of the relationship between Experiment 1 and Experiment 2?
- What is *x* at the given *y*-value?

Research summaries

These passages account for 45% to 55% of the ACT Science Test. Passages present the design, implementation, and conclusion of various scientific experiments. These passages typically contain 5 or 6 questions designed to test one's ability to identify the following:

- What question is the experiment trying to answer?
- What is the researcher's predicted answer to the question?
- How did the researchers test the hypotheses?
- Based on data obtained from the experiments, was the prediction correct?
- What would happen if . . .?
- 2 X 2 Matrix questions: "Yes, because . . ." or "No, because . . ."

Conflicting viewpoints

This section will present a disagreement between two scientists about a specific scientific hypothesis or concept. These passages account for 15% to 20% of the ACT Science Test. The opinions of each researcher are presented in two separate passages. There are two formats for these questions. The test taker must demonstrate an understanding of the content or compare and contrast the main differences between the two opinions. For example:

- Based on the data presented by Scientist 1, which of the following is correct?
- What is the main difference between the conclusions of Expert A and Expert B?

Elements of Science Passages

Short answer questions

Short answer questions are usually one to five words long and require the reader to recall basic facts presented in the passage.

An example answer choice: "Research group #1 finished last."

Long answer questions

Long answer questions are composed of one to three sentences that require the reader to make comparisons, summaries, generalizations, or conclusions about the passage.

An example answer choice: "Scientist 2 disagreed with Scientist 1 on the effects of Bisphenol A pollution and its presumed correlation to birth defects in mice and humans. Scientist 2 proposes that current environmental Bisphenol A levels are not sufficient to cause adverse effects on human health."

Fact questions

Fact questions are the most basic type of question on the test. They ask the reader to recall a specific term, definition, number, or meaning.

An example question/answer: "Which of the following organisms was present in fresh water samples from Pond #1?"

a. Water flea
b. Dragonfly nymph
c. Snail
d. Tadpole

Graphics

There are several different types of graphics used in the ACT Science Test to represent the passage data. There will be at least one of each of the following types included in the test: tables, illustrative diagrams, bar graphs, scatter plots, line graphs, and region graphs. Most ACT Science passages will include two or more of these graphics.

The **illustrative diagram** provides a graphic representation or picture of some process. Questions may address specific details of the process depicted in the graphic. For example, “At which stage of the sliding filament theory of muscle contraction does the physical length of the fibers shorten (and contract)?”

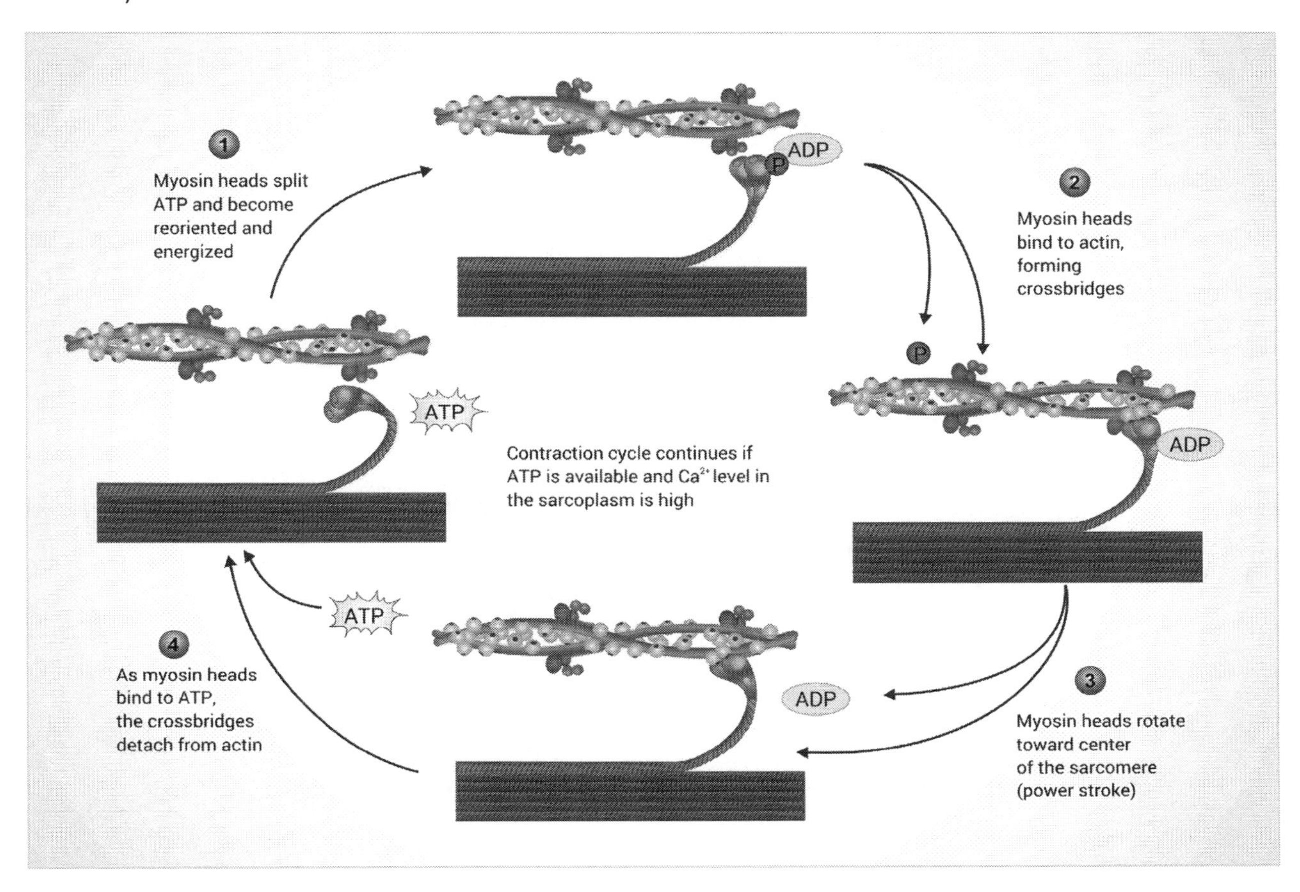

Bar graphs depict the passage data as parallel lines of varying heights. ACT bar graphs will be printed in black and white. Data may be oriented vertically or horizontally. Questions may ask, "During the fall season, in what habitat do bears spend the most time?"

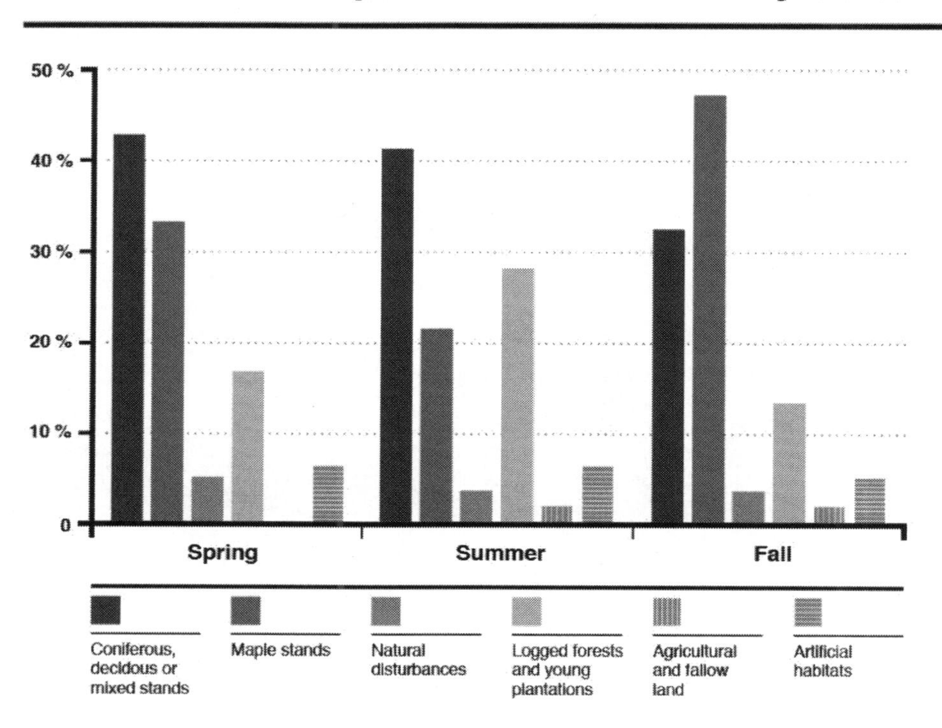

Bar graphs can also be horizontal, like the graph below.

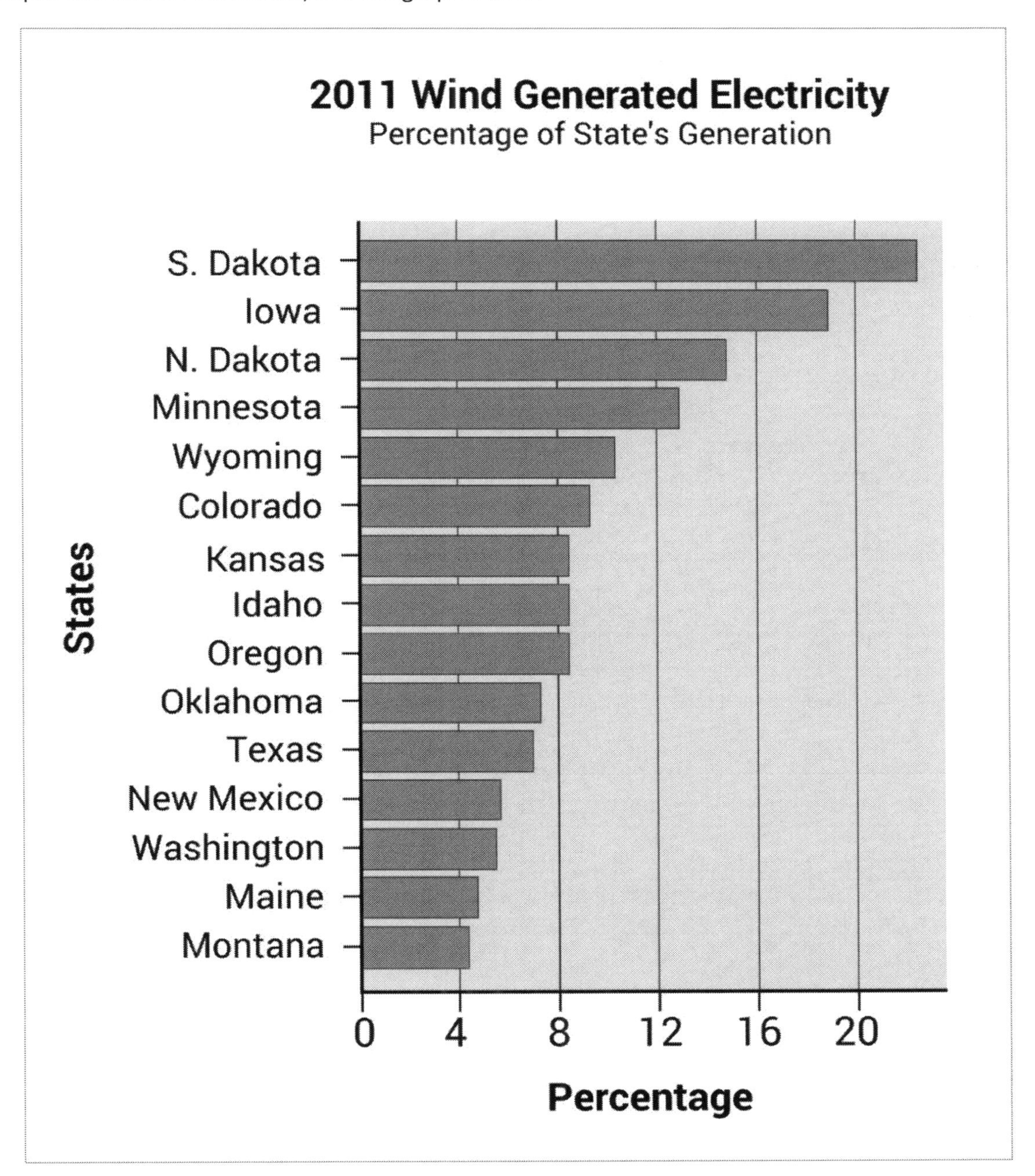

Scatter plots provide a visual representation of the passage data along the x- and y-axes. This representation indicates the nature of the relationship between the two variables. It is important to note that correlation doesn't equal causation. The relationship may be linear, curvilinear, positive, negative, inverse, or there may be no relationship. Questions may ask "What is the relationship between x and y?"

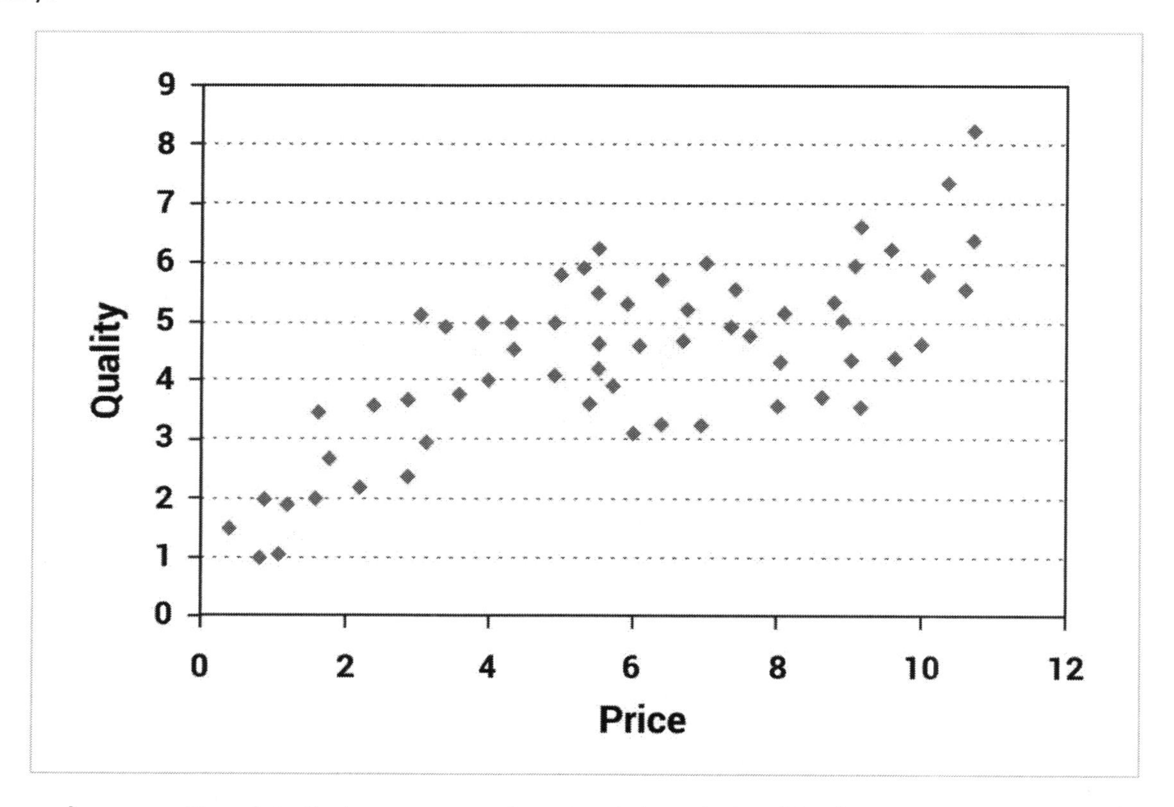

Line graphs are scatter plots that compare and contrast the relationships between two or more data sets. The horizontal axis represents the passage data sets that are compared over time. The vertical axis is the scale for measurement of that data. The scale points are equidistant from one another. There will always be a title for the line graph. Questions related to line graphs might ask," Which of the following conclusions is supported by the provided graph of tropical storms?"

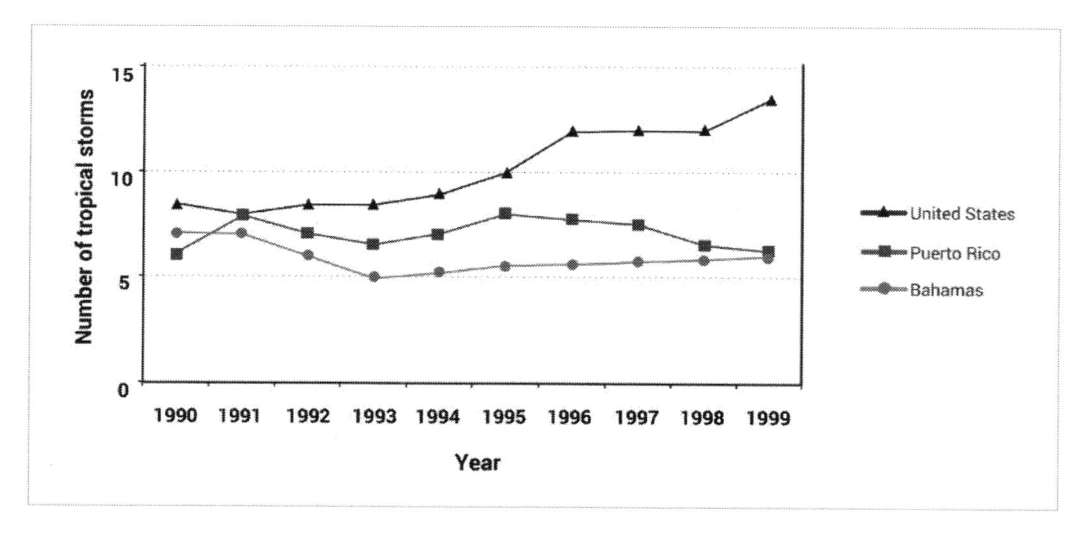

A **region graph** is a visual representation of the passage data set used to display the properties of a given substance under different conditions or at different points in time. Questions relating to this graph may ask, “According to the figure, what is the temperature range associated with liquid nitrogen?”

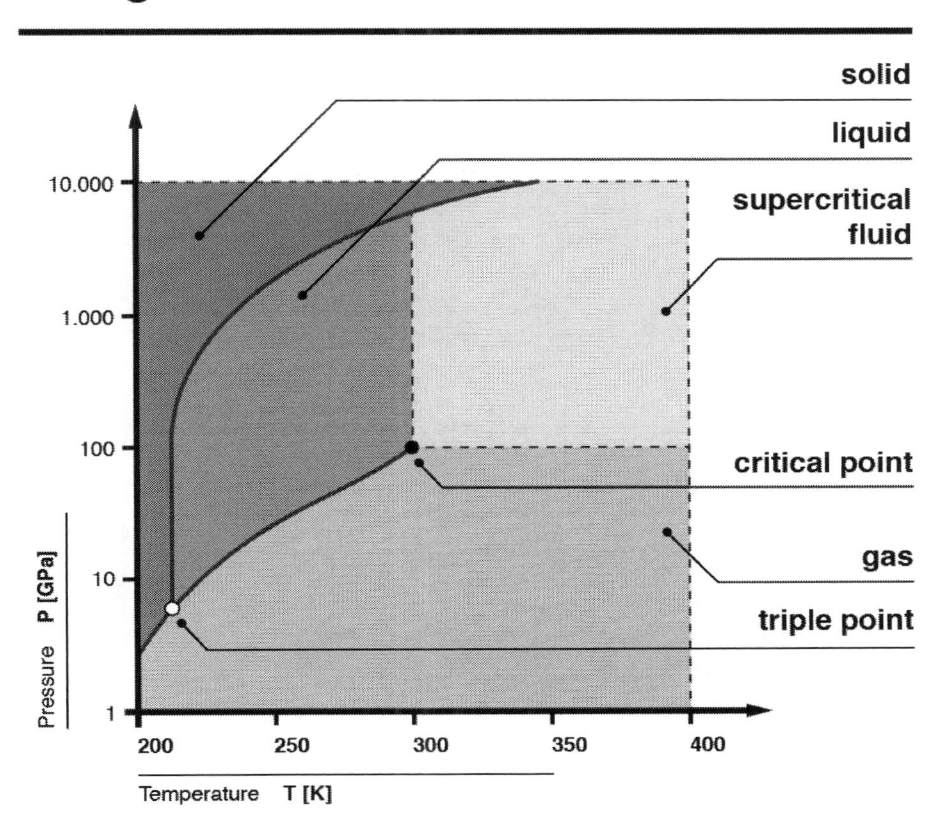

Tables

ACT Science Test tables present passage data sets in tabular form. The **independent variable** is positioned on the left side, while the **dependent variable** is on the right side of the table. The content of ACT Science tables is always discussed in the corresponding passage. Knowledge of all table content isn't required.

Sample Table for Analysis

Data type	Seismic sources							
	Individual faults						Area/volume sources	
	Location	Activity	Length	Dip	Depth	Style	Area	Depth
Geological/Remote Sensing								
Detailed mapping	X	X	X	X		X		
Geomorphic data	X	X	X			X	X	
Quaternary surface rupture	X	X	X			X		
Fault trenching data	X	X		X		X		
Paleoliquefaction data	X	X					X	
Borehole data	X	X		X		X		
Aerial photography	X	X	X					
Low sun-angle photography	X	X	X					
Stellite imagery	X		X				X	
Regional structure	X			X		X	X	
Balanced Cross Section	X			X	X		X	
Geophysical/Geodetic								
Regional potential field data	X		X				X	X
Local potential field data	X		X	X	X	X		
High resolution reflection data	X	X		X		X		
Standard reflexing data	X			X		X		
Deep crustal reflection data	X			X	X		X	X
Tectonic geodetic/strain data	X	X		X	X	X	X	X
Reginal stress data						X	X	
Seismological								
Reflected crustal phase data								X
Pre-instrumental earthquake data	X	X			X	X	X	
Teleseismic earthquake data							X	
Regional network seismicity data	X	X	X	X	X		X	X
Local network seismicity data	X	X	X	X	X			X
Focal mechanism data				X		X		

Answer Choice Elimination Techniques

The ability to quickly and effectively eliminate incorrect answer choices is one of the most important skills to develop before sitting for the exam. The following tips and strategies are designed to help in this effort.

First, test takers should look for answer choices that make absolute "all-or-nothing" statements, as they are usually incorrect answers. Key word clues to incorrect answer choices include: always, never, everyone, no one, all, must, and none. An example of a possible incorrect answer choice is provided below:

> *"The use of antibiotics should be discontinued indefinitely as they are always over-prescribed."*

An example of a possible correct answer choice:

> *"The use of antibiotics should be closely monitored by physician review boards in order to prevent the rise of drug-resistant bacteria."*

Test takers should look for unreasonable, awkward, or irrational answer choices and they should pay special attention to irrelevant information that isn't mentioned in the passage. If the answer choice relates to an idea or position that isn't clearly presented in the passage, it's likely incorrect.

Answer choices that seem too broad or generic and also those that are very narrow or specific are also important to note. For example, if a question relates to a summarization of the passage, test takers should look for answers that encompass a broad representation of the presented ideas, facts, concepts, etc. On the other hand, if a question asks about a specific word, number, data point, etc. the correct answer will likely be more narrow or specific.

Highly Technical Questions

The goal of the ACT Science Test is to measure scientific reasoning skills, not scientific content knowledge. Therefore, candidates shouldn't "cram" for this section using complex scientific concepts from textbooks or online courses. Instead, they should focus on understanding the format of the exam and the manner in which ideas are presented and questions are asked.

> *Example question about two scientists' conflicting viewpoints:*
>
> Scientist 1: Reproductive success in red-tailed hawks has been drastically reduced as a consequence of widespread release of endocrine-disrupting chemicals such as dichloro-diphenyl-trichloroethane (DDT).
>
> Scientist 2: The reproductive fitness of red-tailed hawks is only mildly reduced by the weak effects of DDT, but it is highly sensitive to another class of chemicals called polychlorinated biphenyls (PCBs).

Considering the two viewpoints of the passage, which of the following statements is true?

a. Red-tailed hawks are an endangered species due the effects of over-hunting and the release of toxic chemicals such as DDT and PCBs.
b. PCBs must not be the cause of red-tailed hawk reproductive problems because polychlorinated biphenyls do not affect female reproductive organs.
c. Scientist 2 proposes that PCBs are likely the cause of red-tailed hawk reproductive problems since DDT is only mildly toxic.
d. DDT is a persistent organic pesticide and should be avoided.

Answer A: INCORRECT. This answer is irrational. Neither scientist mentioned anything about hunting or red-tailed hawks being endangered.

Answer B: INCORRECT. This answer is also irrational. Neither scientist proposes that PCBs do not affect female reproductive organs.

Answer C: CORRECT. This answer makes the most sense. It's true that Scientist 2 believes that PCBs cause reproductive problems in red-tailed hawks because DDT is only mildly toxic.

Answer D: INCORRECT. This answer is very narrow and specific. The phrase "persistent organic pesticide" isn't mentioned at all in the passage.

Again, the reader didn't need to know anything about dichloro-diphenyl-trichloroethane, polychlorinated biphenyls, red-tailed hawks, reproductive biology, or endocrine-disrupting chemicals to answer the question correctly. Instead, the test taker must consider both viewpoints and make comparisons or generalizations between the two.

Time Management

Using time efficiently is a critical factor for success on the ACT Science Test. Test takers may find the following strategies to be helpful when preparing for the exam:

- There are 40 questions and only 35 minutes, which leaves about 50 seconds for each question. This may seem daunting; however, it's important to consider that multiple questions apply to a single passage, graph, etc. Therefore, it may be possible to answer some in less than 50 seconds, leaving extra time for more demanding questions.
- It's important to read the entire passage before answering the questions and to consider all facts, conflicting viewpoints, hypotheses, conclusions, opinions, etc.
- Be sure to review only the information presented in the passages, as the questions will relate solely to this content.
- Don't get stuck on difficult questions. The exam is designed with a mix of easy and hard questions. Make sure to get all the easy ones answered, and then go back to focus on more difficult questions.
- Try to quickly eliminate answer choices that are obviously incorrect.

General Tips for Success

Before taking the exam, successful test takers practice answering questions from all of the different formats. Being able to quickly identify what one is being asked to do saves valuable time.

Test takers should read each passage completely and make notes or underline important information, such as:

- What is the hypothesis of an experiment?
- What is the result of the experiment?
- What is the experimental question?
- What information led to the researcher's viewpoint?
- What is the main difference between the two experiments?
- What is the main difference of opinion between two researchers?

Test takers should be mindful of the time. There are 40 questions that must be completed in 35 minutes. It is recommended that test takers quickly look at the questions before reading the associated passage. This will provide a general idea of which parts of the passage to read carefully. Test takers should answer all of the questions.

ACT Science Practice Test #1

Passage 1

Questions 1–5 pertain to Passage 1:

Predators are animals that eat other animals. Prey are animals that are eaten by a predator. Predators and prey have a distinct relationship. Predators rely on the prey population for food and nutrition. They evolve physically to catch their prey. For example, they develop a keen sense of sight, smell, or hearing. They may also be able to run very fast or camouflage to their environment in order to sneak up on their prey. Likewise, the prey population may develop these features to escape and hide from their predators. As predators catch more prey, the prey population dwindles. With fewer prey to catch, the predator population also dwindles. This happens in a cyclical manner over time.

Figure 1 below shows the cyclical population growth in a predator-prey relationship.

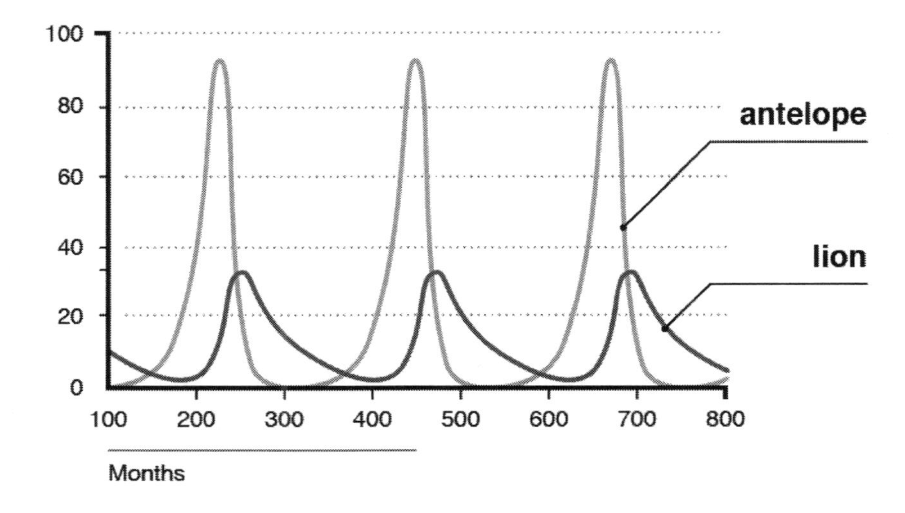

Figure 2 below shows a predator-prey cycle in a circular picture diagram

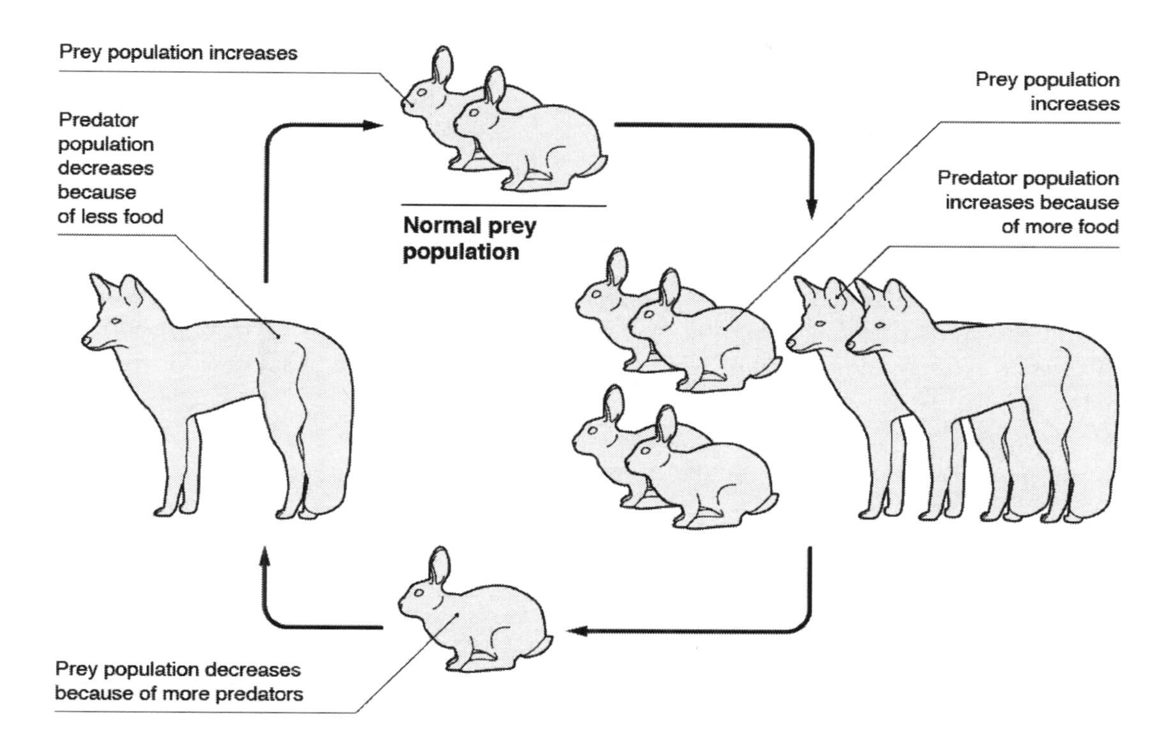

1. Looking at Figure 1, approximately how long is one cycle of the prey population, which includes the population being low, reaching a peak, and then becoming low again?
    a. 200 months
    b. 100 months
    c. 800 months
    d. 400 months

2. In Figure 2, which animal is the predator?
    a. Both the fox and rabbit
    b. Fox only
    c. Rabbit only
    d. Neither the fox nor the rabbit

3. What causes the predator population to decrease?
    a. When there's an increase in the prey population
    b. When winter arrives
    c. When the prey start attacking the predators
    d. When there are fewer prey to find

4. What causes the prey population to increase?
    a. When the predator population decreases, so more prey survive and reproduce.
    b. When there's an increase in the predator population
    c. When there's more sunlight
    d. The prey population always remains the same size.

5. Which is NOT a feature that a prey population can develop to hide from their predator?
   a. Keen sense of smell
   b. Camouflage ability
   c. A loud voice
   d. Keen sense of hearing

Passage 2

Questions 6–10 pertain to Passage 2:

Greenhouses are glass structures that people grow plants in. They allow plants to survive and grow even in the cold winter months by providing light and trapping warm air inside. Light is allowed in through the clear glass walls and roof. Warm air comes in as sunlight through the glass roof. The sunlight is converted into heat, or infrared energy, by the surfaces inside the greenhouse. This heat energy then takes longer to pass back through the glass surfaces and causes the interior of the greenhouse to feel warmer than the outside climate.

Plants may grow better inside a greenhouse versus outside for several reasons. There is more control of the temperature and humidity of the environment inside the greenhouse. The carbon dioxide produced by plants is trapped inside the greenhouse and can increase the rate of photosynthesis of the plants. There are also fewer pests and diseases inside the greenhouse.

Figure 1 below shows how a greenhouse works.

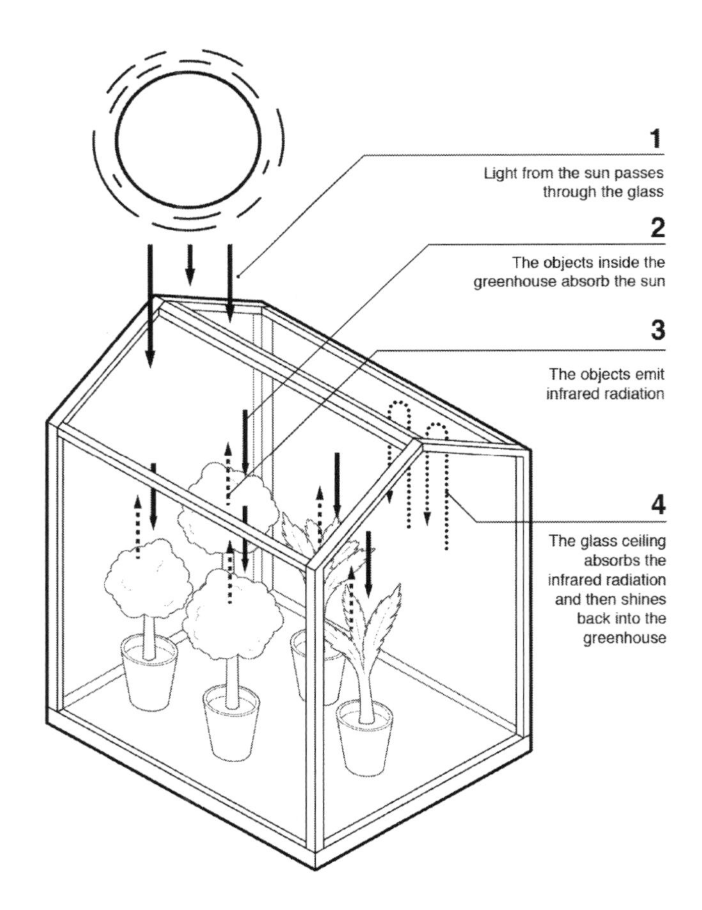

Scientist A wants to compare how a tomato plant grows inside a greenhouse versus outside a greenhouse.

Figure 2 below shows a graph of her results over 3 months.

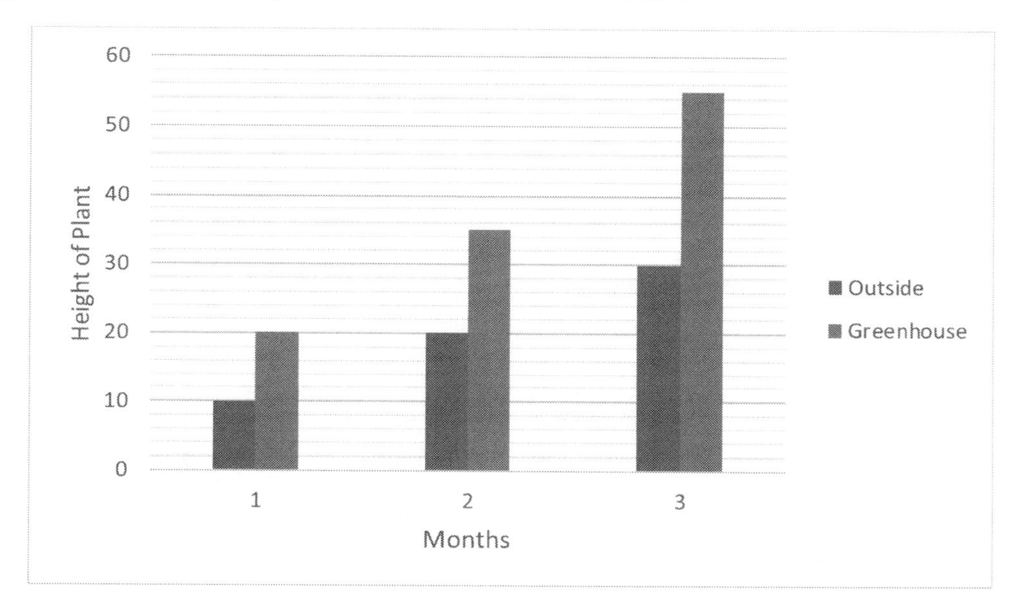

6. Looking at Figure 1, what gets trapped inside the greenhouse that helps plants grow?
   a. Short-wavelength IR
   b. Long-wavelength IR
   c. Cold air
   d. Water

7. Which plant grew taller from Scientist A's experiment?
   a. Outside
   b. Both grew to the same height.
   c. They both remained the same height for 3 months.
   d. Greenhouse

8. What gets converted to heat inside a greenhouse?
   a. Water
   b. Sunlight
   c. Plants
   d. Oxygen

9. What type of wavelength moves through the greenhouse glass easily according to Figure 1?
   a. Short-wavelength IR
   b. Oxygen
   c. Carbon dioxide
   d. Long-wavelength IR

10. What is one reason that plants may grow better inside a greenhouse?
    a. Colder air
    b. Less photosynthesis occurs in the greenhouse
    c. Fewer pests
    d. Less sunlight comes into the greenhouse

Passage 3

Questions 11–15 pertain to Passage 3:

In chemistry, a titration is a method that is used to determine the concentration of an unknown solution. Generally, a known volume of a solution of known concentration is mixed with the unknown solution. Once the reaction of the two solutions has been completed, the concentration of the unknown solution can be calculated. When acids and bases are titrated, the progress of the reaction is monitored by changes in the pH of the known solution. The equivalence point is when just enough of the unknown solution has been added to neutralize the known solution. A color reaction may also occur so that with the drop of solution that causes complete neutralization, the solution turns bright pink, for example. For acids that only have one proton, usually a hydrogen atom, the halfway point between the beginning of the curve and the equivalence point is where the amount of acid and base are equal in the solution. At this point, the pH is equal to the pK_a, or the acid dissociation constant.

Figure 1 below shows a general titration curve of a strong acid with a strong base.

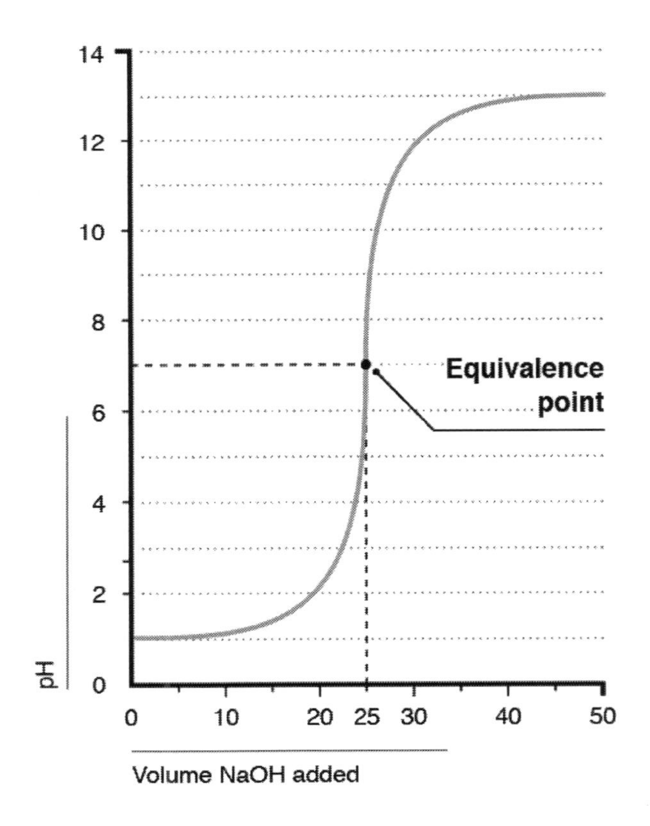

Figure 2 below shows the chemical reaction of a strong acid with a strong base.

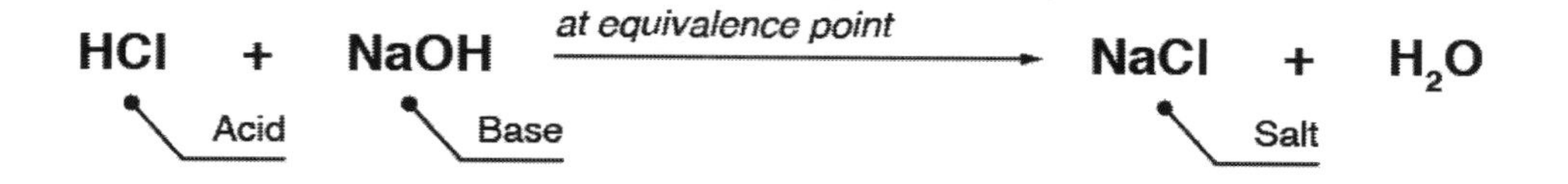

Figure 3 shows the titration curve for acetic acid.

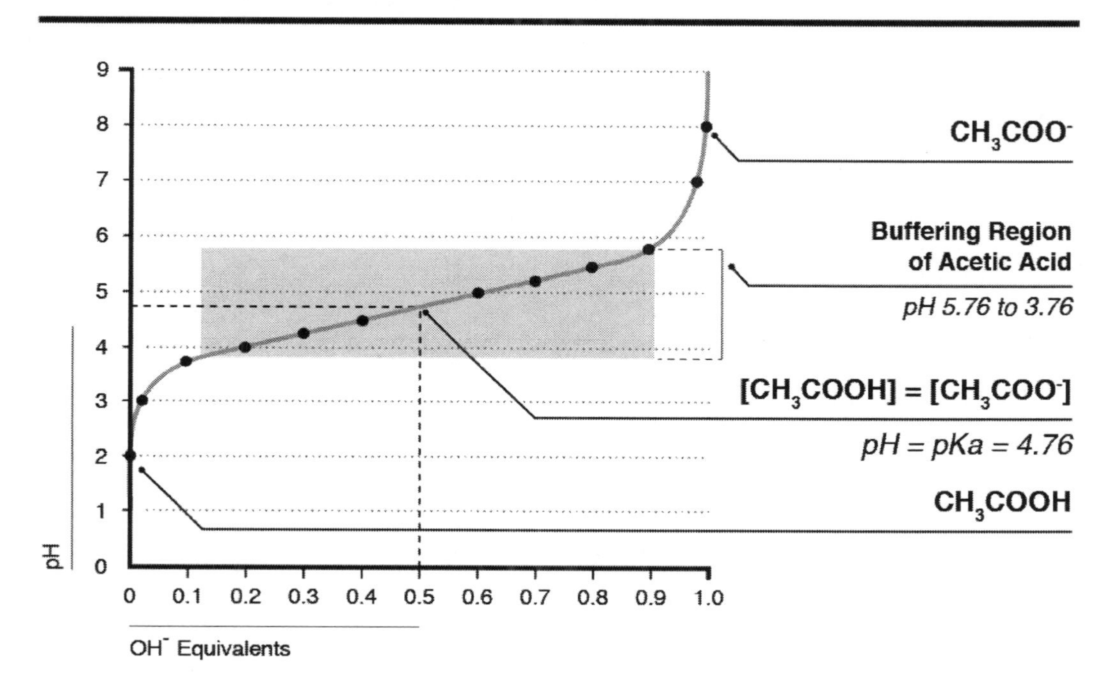

11. How much NaOH is added to the HCl solution to reach the equivalence point in Figure 1?
    a. 10
    b. 40
    c. 50
    d. 25

12. What is the acid dissociation constant of the titration curve in the Figure 3?
    a. 4.21
    b. 3.50
    c. 4.76
    d. 6.52

13. What is the pH of the acetic acid before the titration has started in Figure 3?
    a. 2
    b. 4.76
    c. 7
    d. 6

14. What is one of the products of the chemical equation in Figure 2?
    a. HCl
    b. NaCl
    c. NaOH
    d. Cl^-

15. How would you describe the solution at the equivalence point in Figure 1?
    a. Neutral
    b. Acidic
    c. Basic
    d. Unknown

Passage 4

Questions 16–20 pertain to Passage 4:

> The heart is a muscle that is responsible for pumping blood through the body. It is divided into four chambers: the right atrium, right ventricle, left atrium, and left ventricle. Blood enters the atria and is then pumped into the ventricles below them. There is a valve between the atria and ventricles that prevents the blood from flowing back into the atria. The valve between the right atrium and ventricle has three folds whereas the valve between the left atrium and ventricle has two folds. Arteries carry oxygen-rich blood away from the heart to the body. Veins carry oxygen-poor blood from the body back to the heart. From there, the blood gets pumped to the lungs to get re-oxygenated and then back to the heart before circulating to the body. The heart beats every second of the day. For an adult, the normal heartrate is between 60 and 100 beats per minute. For a child, a normal heartrate is between 90 and 120 beats per minute.

Figure 1 below shows how blood gets pumped through the body.

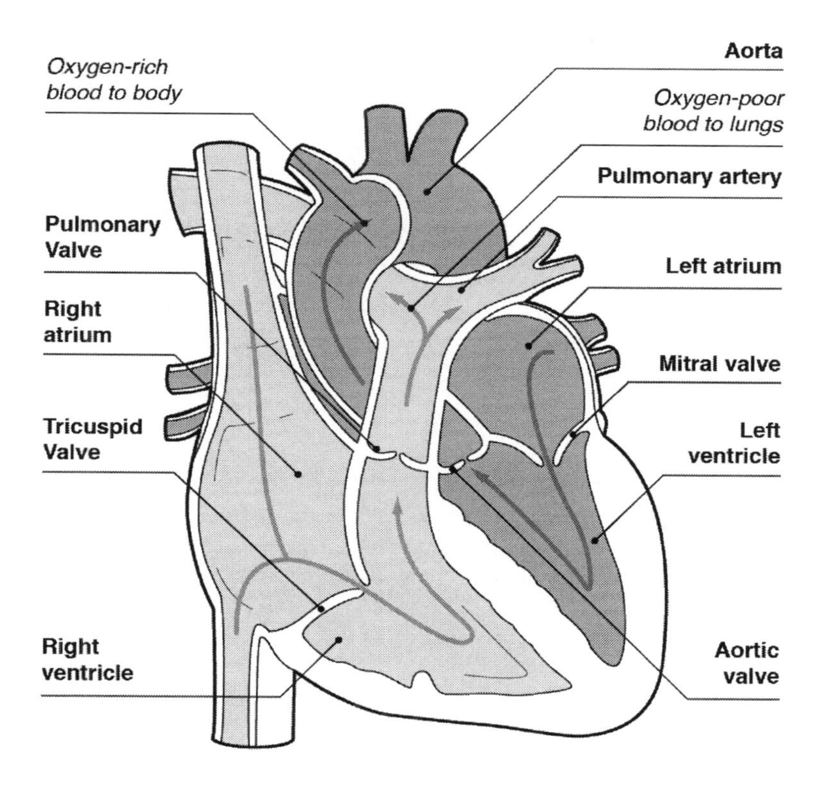

16. Where is the oxygen-poor blood pumped to before returning to the heart to get circulated to the rest of the body?
   a. Lungs
   b. Brain
   c. Stomach
   d. Kidney

17. Which heart valve has two folds?
   a. Pulmonary artery
   b. Tricuspid valve
   c. Mitral valve
   d. Aorta

18. If the aorta contains oxygen-rich blood, what type of vessel is it?
   a. Vein
   b. Pulmonary
   c. Airway
   d. Artery

19. Which heartrate (beats per minute) would be considered normal for a child during resting conditions?
    a. 85
    b. 125
    c. 100
    d. 60

20. The aorta is the artery that breaks into smaller vessels to transport blood to the rest of the body. Looking at the figure, which is the final chamber that the blood flows through before entering the aorta?
    a. Right ventricle
    b. Left ventricle
    c. Right atrium
    d. Left atrium

Passage 5

Questions 21–25 pertain to Passage 5:

> There are three types of rocks: sedimentary, metamorphic, and igneous. Sedimentary rock is formed from sediment, such as sand, shells, and pebbles. The sediment gathers together and hardens over time. It is generally soft and breaks apart easily. This is the only type of rock that contains fossils, which are the remains of animals and plants that lived a long time ago. Metamorphic rock forms under the surface of the earth due to changes in heat and pressure. These rocks usually have ribbon-like layers and may contain shiny crystals. Igneous rock forms when molten rock, or magma, cools and hardens. An example of molten rock is lava, which escapes from an erupting volcano. This type of rock looks shiny and glasslike.

Figure 1 below is a chart of different types of rocks.

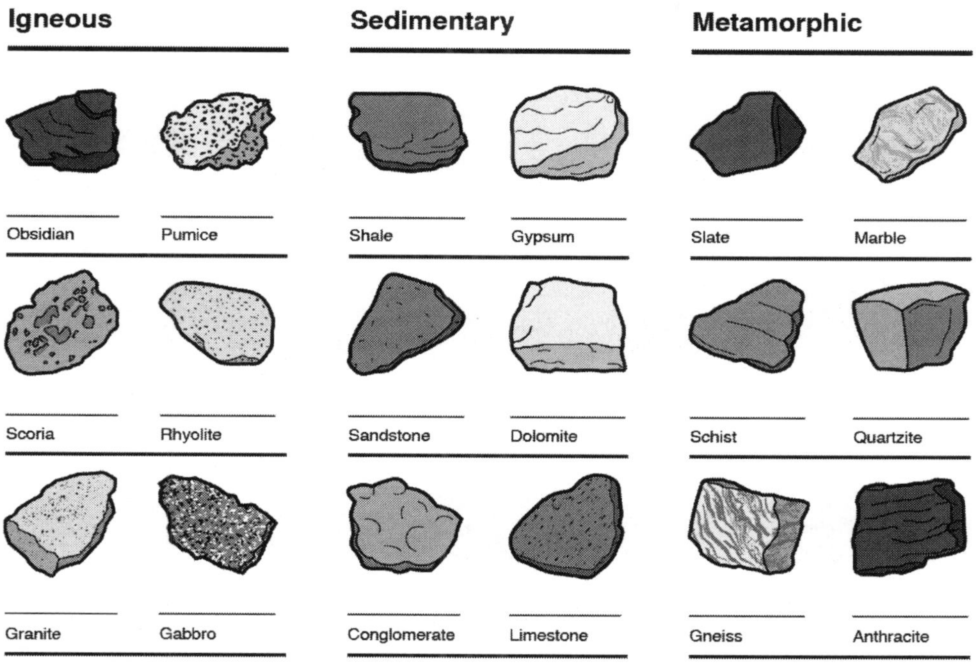

Igneous		Sedimentary		Metamorphic	
Obsidian	Pumice	Shale	Gypsum	Slate	Marble
Scoria	Rhyolite	Sandstone	Dolomite	Schist	Quartzite
Granite	Gabbro	Conglomerate	Limestone	Gneiss	Anthracite

21. A volcano erupts and lava comes out and hardens once it is cooled. What type of rock is formed?
    A. Sedimentary
    B. Metamorphic
    C. Igneous
    D. Lava does not cool

22. Scientist A found a piece of granite rock, as seen in Figure 1. What type of rock is it?
    a. Igneous
    b. Metamorphic
    c. Sedimentary
    d. Fossil

23. Which type of rock could a fossil be found in?
    a. Igneous
    b. Bone
    c. Metamorphic
    d. Sedimentary

24. Which is an example of a metamorphic rock in the figure?
    a. Sandstone
    b. Slate
    c. Granite
    d. Limestone

25. What type of rock would most likely be formed at and found on the beach?
   a. Sedimentary
   b. Shells
   c. Igneous
   d. Metamorphic

Passage 6

Questions 26–30 pertain to Passage 6:

> The greenhouse effect is a natural process that warms the Earth's surface, similar to what occurs in a greenhouse meant to grow plants. Solar energy reaches the Earth's atmosphere and warms the air and land. Some of the energy is absorbed by the greenhouse gases found in the Earth's atmosphere and by the land, and the rest is reflected back into space. Greenhouse gases include water vapor, carbon dioxide, methane, nitrous oxide, and chlorofluorocarbons. In recent decades, human activity has increased the amount of greenhouse gases present in the Earth's atmosphere, which has created a warmer atmosphere than normal and increased the Earth's temperature.
>
> Figure 1 below shows the process of the greenhouse effect.

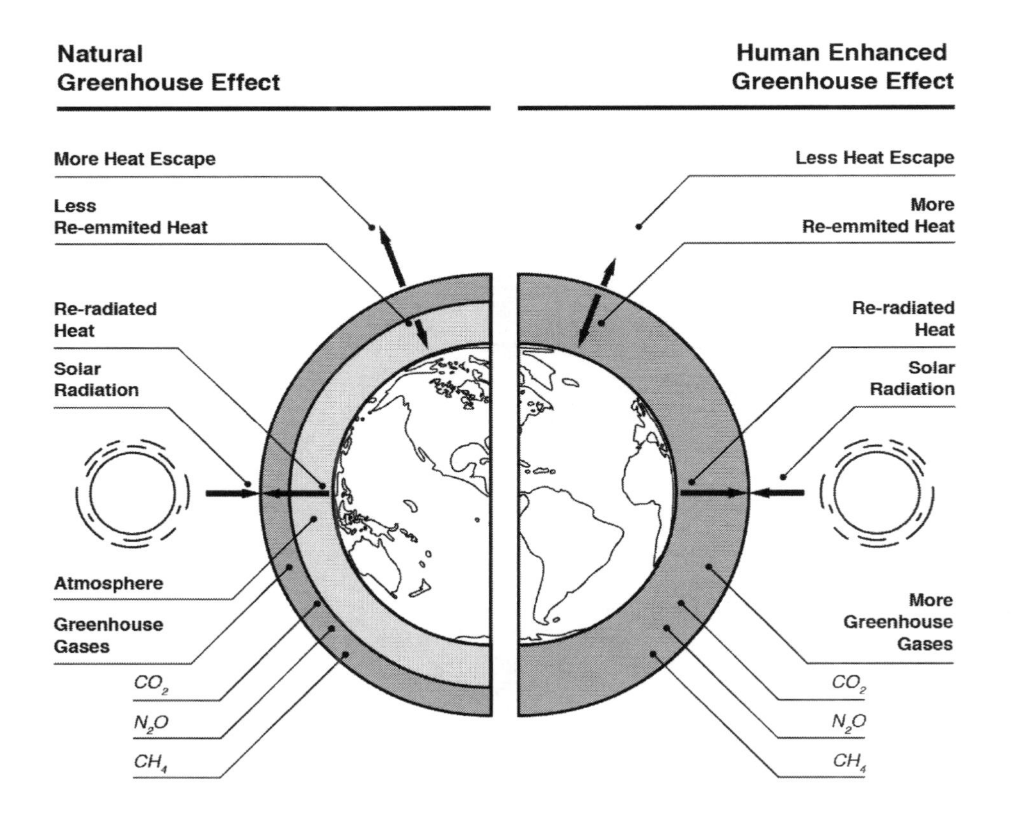

Figure 2 below describes the greenhouse gases that are produced from human activity.

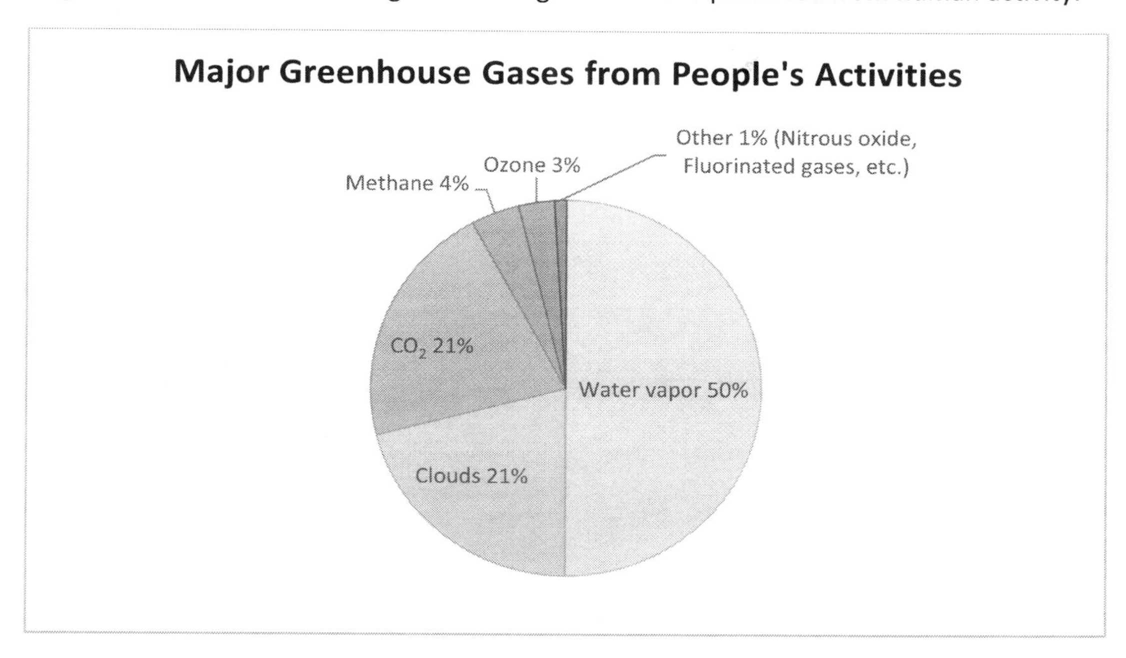

Figure 3 below describes the human activities that produce carbon dioxide.

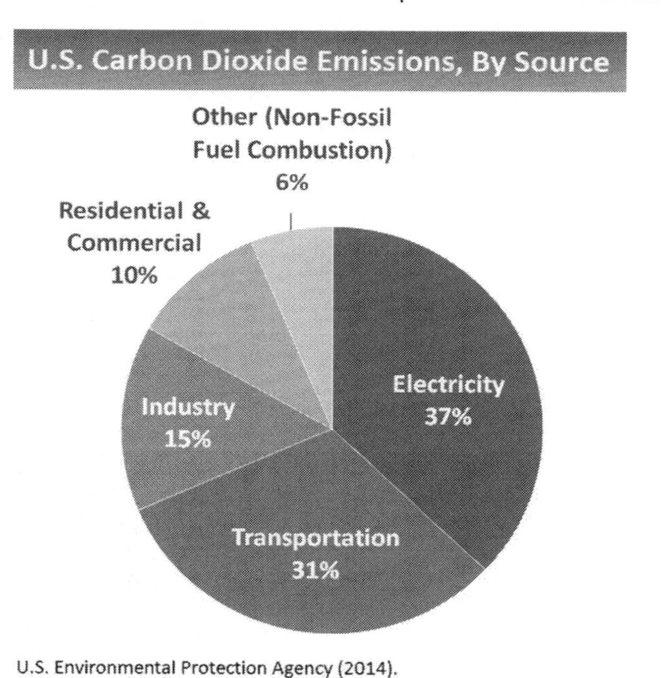

26. Which is NOT an example of a greenhouse gas?
   a. Methane
   b. Carbon dioxide
   c. Nitrous oxide
   d. Helium gas

27. Looking at Figure 3, what could a person do to decrease how much carbon dioxide they produce?
    a. Leave lights on all the time
    b. Walk instead of drive a car
    c. Always drive in their own car everywhere
    d. Leave the television on all the time

28. Looking at Figure 2, which is the second highest greenhouse gas produced by human activity?
    a. Methane
    b. Fluorinated gases
    c. Nitrous oxide
    d. Carbon dioxide

29. Looking at Figure 1, what gets increasingly trapped in the Earth's atmosphere with increased human activity?
    a. Space
    b. The Sun
    c. Heat
    d. Water vapor

30. What type of charts are found in Figures 2 and 3?
    a. Scatter plots
    b. Line graphs
    c. Bar graphs
    d. Pie charts

Passage 7

Questions 31–35 pertain to Passage 7:

> A meteorologist uses many different tools to predict the weather. They study the atmosphere and changes that are occurring to predict what the weather will be like in the future. Listed below are some of the tools that a meteorologist uses:
>
> - Thermometer: measures air temperature
> - Barometer: measures air pressure
> - Rain gauge: measures rainfall over a specific time
> - Anemometer: measures air speed
> - Wind vane: shows which direction the wind is blowing

Figure 1 below shows data that is collected by a meteorologist.

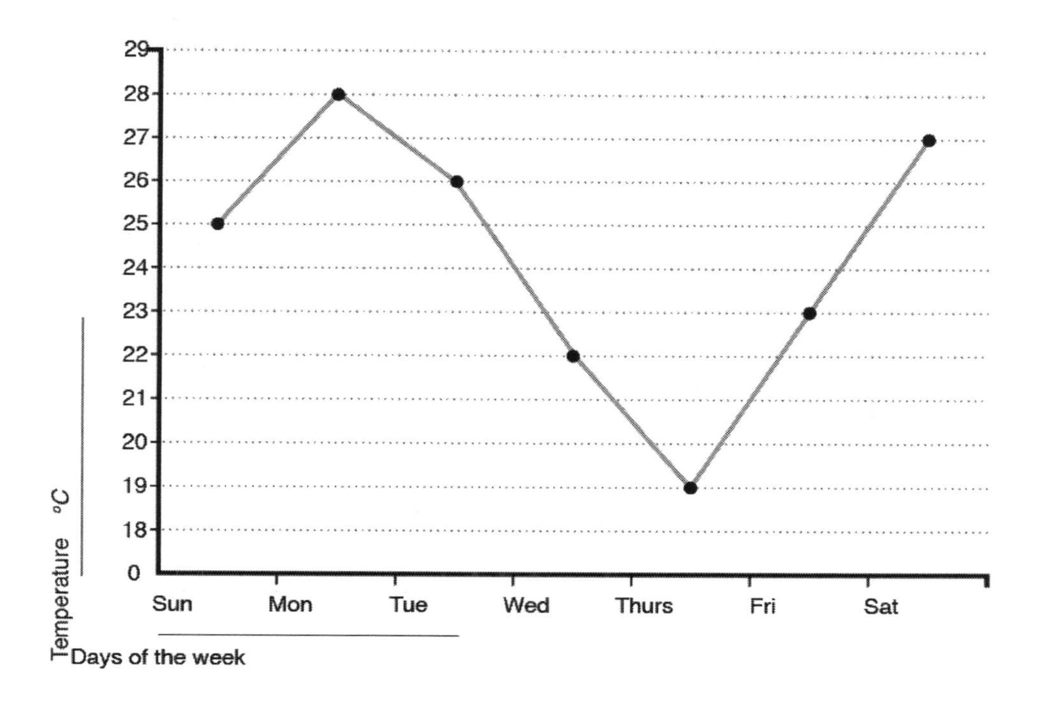

Figure 2 below shows data collected from a rain gauge.

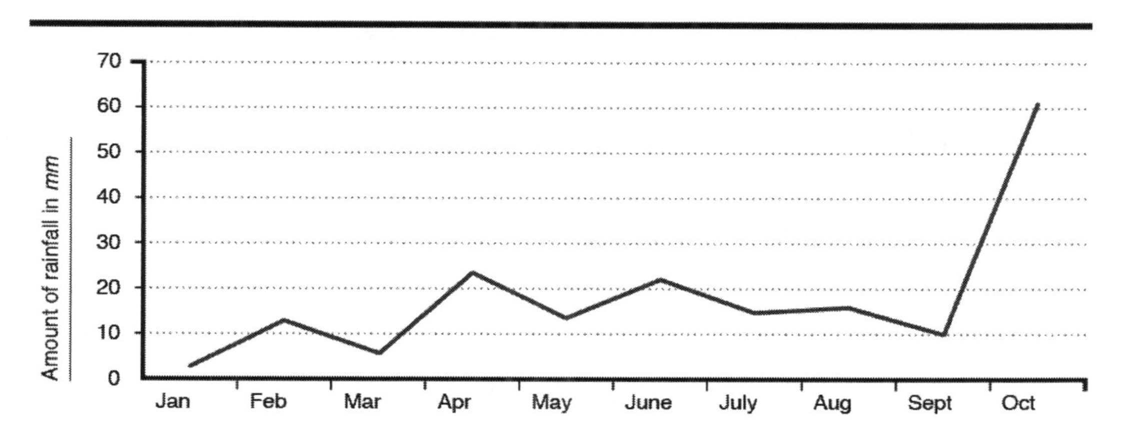

31. What tool would a meteorologist use to find out how fast the wind is blowing?
    a. Anemometer
    b. Barometer
    c. Thermometer
    d. Wind vane

32. What tool was used to collect the data shown in Figure 1?
   a. Rain gauge
   b. Thermometer
   c. Barometer
   d. Anemometer

33. The wind vane is pointing north. What does this tell us?
   a. Wind is blowing in an eastern direction.
   b. A storm is coming.
   c. The wind is blowing in a northern direction.
   d. The wind is blowing in a southern direction.

34. Looking at Figure 2, which month had the lowest rainfall?
   a. January
   b. April
   c. September
   d. October

35. Looking at Figure 2, what was the approximate amount of rain that fell in June?
   a. 0 mm
   b. 10 mm
   c. 50 mm
   d. 20 mm

Passage 8

Questions 36–40 pertain to Passage 8:

> Cells are the smallest functional unit of living organisms. Organisms can be single-celled or multicellular. Each cell contains organelles that are responsible for distinct functions and are essential for the organism's life. Plants and animals have different necessities for generating energy and nutrients. Their cells are similar but also have unique features.

Figure 1 below is a depiction of the organelles in an animal cell.

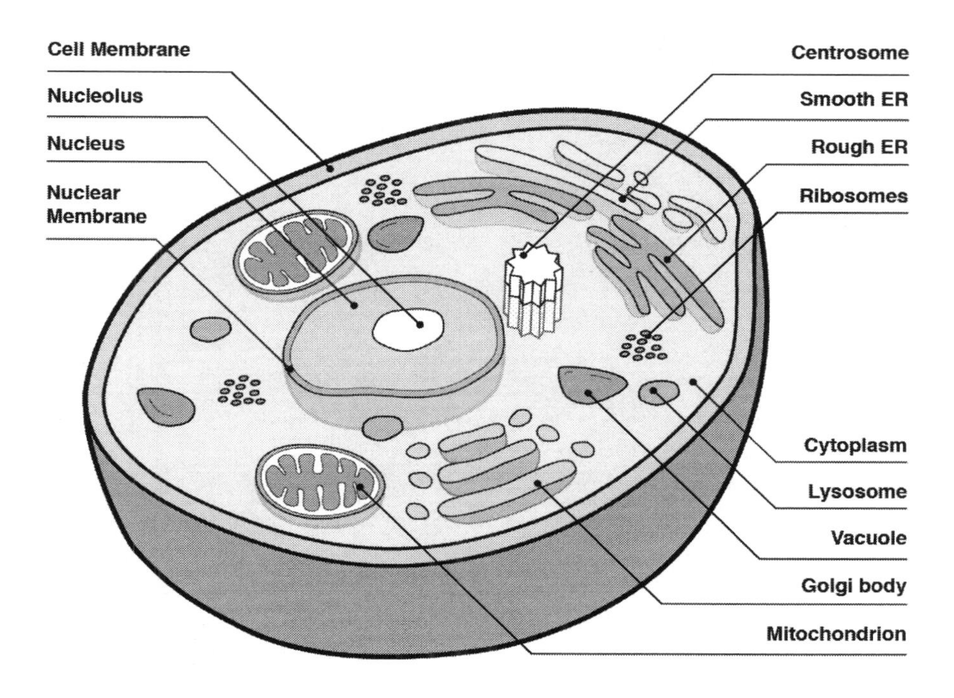

Figure 2 below depicts the organelles of a plant cell.

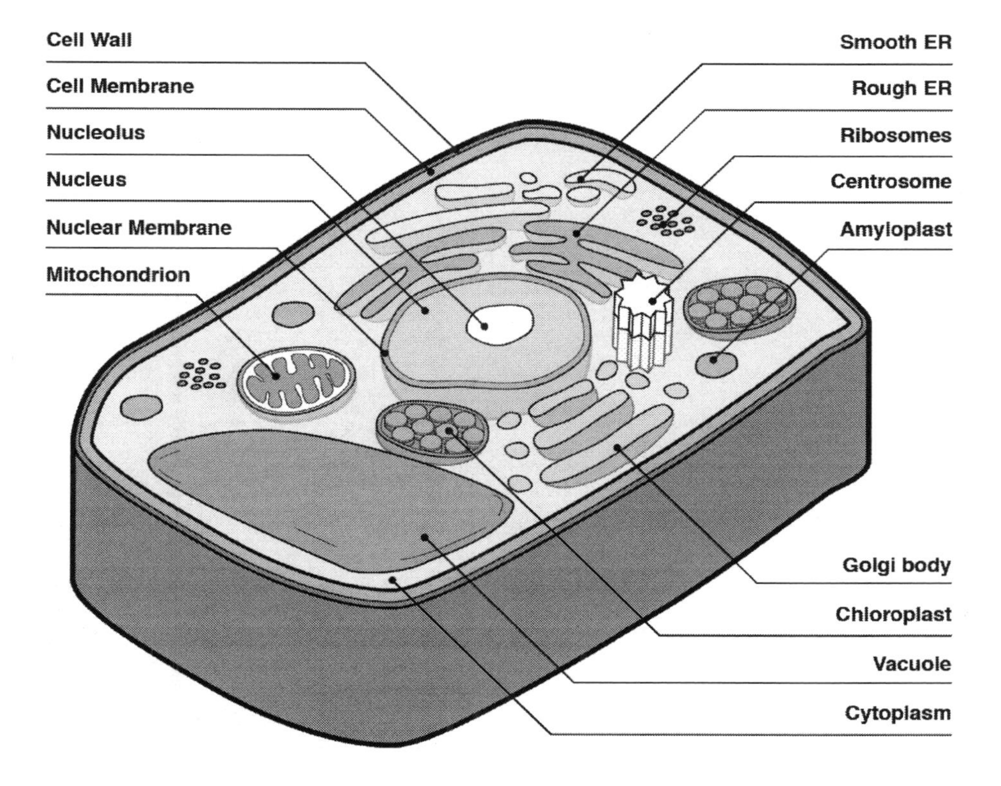

Figure 3 below describes the function of many important organelles.

Cell Organelle	Function
Cell wall	*(Plants only)* Maintains the shape of the cell and is a protective barrier for the internal contents of the cell.
Chloroplasts	*(Plants only)* Site of photosynthesis, which converts sunlight energy to glucose storage energy.
Nucleus	*(Plants and Animals)* Contains the cell's DNA.
Ribosomes	*(Plants and Animals)* Puts together long chains of amino acids to build proteins. Smallest organelle in the cell.
Mitochondria	*(Plants and Animals)* The powerhouse of the cell. Converts the stored glucose energy to ATP energy, which drives forward almost all of the cell's reactions.
Cell membrane	*(Plants and Animals)* Regulates what molecules can move in and out of the cell. Made of a phospholipid bilayer.
Cytoplasm	*(Plants and Animals)* The liquid that fills the inside of the cell.
Vacuole	*(Plants and Animals)* A membranous sac that encloses anything in the cell that needs to be kept separate, such as food and water.
Golgi Body	*(Plants and Animals)* Receives products produced by the endoplasmic reticulum (ER) and adds final changes to them.
Lysosomes	*(Plants and Animals)* A membranous sac that is full of digestive juices. Breaks down larger molecules into smaller parts so that they can be used to build new parts of the cell.
Rough endoplasmic reticulum (ER)	*(Plants and Animals)* A large folded membrane that is covered with ribosomes. Helps fold and modify the proteins built by the ribosomes before sending them to the Golgi body.
Smooth endoplasmic reticulum (ER)	*(Plants and Animals)* A large folded membrane that puts together lipids.
Microtubules and microfilaments	*(Plants and Animals)* Long tubes that allow the cell to move and provide an internal structure of support for the cell.

36. Which is an organelle found in a plant cell but not an animal cell?
    a. Mitochondria
    b. Chloroplast
    c. Golgi body
    d. Nucleus

37. Where is the nucleolus located in both plant and animal cells?
    a. Near the chloroplast
    b. Inside the mitochondria
    c. Inside the nucleus
    d. Attached to the cell membrane

38. Which organelle is responsible for generating energy for the cell and is referred to as the powerhouse of the cell?
    a. Mitochondria
    b. Nucleus
    c. Ribosomes
    d. Cell wall

39. What does the cell membrane do?
    a. Builds proteins
    b. Breaks down large molecules
    c. Contains the cell's DNA
    d. Controls which molecules are allowed in and out of the cell

40. What are chloroplasts responsible for in plant cells?
    a. Maintaining the cell's shape
    b. Containing the cell's DNA
    c. Converting energy from sunlight to glucose
    d. Building proteins

Answer Explanations #1

1. A: One cycle takes 200 months. It starts with the population being low, rising and reaching a peak, and then falling again. It takes 100 months to reach the peak, Choice *B*. Four cycles could be completed in 800 months, Choice *C*, and two cycles could be completed in 400 months, Choice *D*.

2. B: Looking at the Figure 2, the fox is the predator. When the diagram notes that the predator population decreases on the left side, there is only one fox left. As it increases, as noted on the right side, there are two foxes drawn. Foxes are also much larger than rabbits and would be able to catch them much easier than the other way around.

3. D: When the prey population decreases, the predators have less food, i.e. prey, to feed on. This causes the predator population to dwindle. An increase in the prey population, Choice *A*, would actually increase the predator population because they would have more food, which would lengthen survival and increase reproduction. Seasons do not affect the predator population in this situation, Choice *B*. Generally, prey do not have the ability to attack their predators, Choice *C*, due to physical constraints, such as differences in size.

4. A: When the predator population decreases, the rate of survival of the prey population increases and they can then also reproduce more. An increase in the predator population, Choice *B*, would cause the prey population to decrease. Weather and amount of sunlight, Choice *C*, does not affect the growth of the prey population. The prey population is cyclical and does not remain the same size, Choice *D*.

5. C: Prey populations can develop different features to try and hide from and escape the predator population. The features help them blend into their environment, such as Choice *B*, or help them identify predators early and quickly, Choices *A* and *D*. Choice *C* would just allow the predators to hear the prey easily.

6. B: Sunlight comes into the greenhouse as short-wavelength IR. As it is absorbed by surfaces in the greenhouse, it is converted to long-wavelength IR. The long-wavelength IR gets trapped inside the greenhouse and bounces off the surfaces and glass and remains inside the greenhouse. Since short-wavelength IR can enter the greenhouse, it also has the ability to leave the greenhouse, making Choice *A* incorrect. The greenhouse feels warmer, not cooler, than outside, so Choice *C* is incorrect. Water is not involved in the reaction noted in Figure 1, so Choice *D* is also incorrect.

7. D: Looking at the graph in Figure 2, the greenhouse plant grew taller than the outside plant. The bars representing the greenhouse plant are taller at 3 time points that Scientist A measured. Greenhouses trap sunlight, warm air, and gases, such as CO_2 inside the greenhouse, so plants have an increased rate of photosynthesis, allowing them to grow faster. The plants are also protected from pests inside the greenhouse.

8. B: Sunlight enters the greenhouse as short-wavelength IR and get converted to long-wavelength IR. This process also gives off heat and makes the greenhouse feel warmer than the outside climate. Water and oxygen, Choices *A* and *D*, are not involved in this reaction. The plants remain the same and do not get converted into anything else, Choice *C*.

9. A: Short-wavelength IR enters the greenhouse in the form of sunlight. It can pass easily through the glass and can therefore pass easily back out to the outside environment. The long-wavelength IR, Choice *D*, gets trapped inside the greenhouse.

10. C: The plants inside a greenhouse are protected from many pests that can be found in the outside environment. The air is warmer in the greenhouse, so Choice *A* is incorrect. More photosynthesis occurs because of the increased sunlight energy that stays in the greenhouse, making Choices *B* and *D* incorrect.

11. D: The equivalence point occurs when just enough of the unknown solution is added to completely neutralize the known solution. In Figure 1, at the halfway point of the curve, the equivalence point is when 25 volumes of NaOH have been added to the solution. The pH is 7 at this point also, which is a neutralization of the HCl, strong acid.

12. C: The acid dissociation constant is the pK_a of the solution. It is found at the halfway point between the beginning of the curve and the equivalence point, where the solution would have a pH of 7 and be completely neutralized. In Figure 3, it is marked as 4.76.

13. B: Looking at Figure 3, the vertical axis on the left side has information about the pH of the solution. The horizontal axis at the bottom has information about how much basic solution containing OH- is being added to the acetic acid. When the OH^- is at 0, and none has been added yet, the pH of the acetic acid is marked as 2.

14. B: Looking at the chemical equation in Figure 2, the reactants are on the left side and the products are on the right side. HCl and NaOH, Choices *A* and *C*, are the reactants of the equation. NaCl is the salt that is formed as one of the products of the reaction. The chloride ion, Choice *D*, is not formed in this reaction.

15. A: The equivalence point occurs in all titration reactions when the solution is neutralized. If an acid and base are being titrated, the solution is no longer acidic or basic, Choices *B* and *C*. It reaches a pH of 7 and is considered neutral.

16. A: Oxygen-poor blood is pumped to the lungs before returning to the heart. Oxygen is transferred from the airways of the lungs into the blood. The blood becomes rich with oxygen and then returns to the heart so that it can bring oxygen and nutrients to other organs of the body, such as the brain, stomach, and kidney, Choices *B*, *C*, and *D*.

17. C: The mitral valve has two folds. The tricuspid valve, Choice *B*, has three folds. The valve between the left atrium and ventricle has two valves, as was noted in the descriptive passage. Correlating this information to Figure 1, the name of the valve between these two chambers is the mitral valve. Choices *A* and *D* are vessels that carry blood through them and are not names of valves.

18. D: Arteries carry oxygen-rich blood away from the heart to the rest of the body. The aorta is the largest artery in the body. Veins, Choice *A*, carry oxygen-poor blood to the heart and lungs. Airway, Choice *C*, is found in the respiratory system and carries air in and out of the body.

19. C: A normal heartrate for a child is between 90 and 120 beats per minute. Choice *C*, 100 beats per minute, falls within this range. Choices *A*, *B*, and *D* are within the normal range for an adult but not for a child. Children's hearts pump blood faster than adults' hearts through the body.

20. B: Looking at Figure 1 and following the red arrow in the aorta backwards, the left ventricle is where the blood is coming from directly before it enters the aorta. Blood flows from the atria to the ventricles, so it enters the left atrium before the left ventricle and then the aorta, Choice *D*. Oxygen-poor blood is on the right side of the heart and flows from the right atrium to the right ventricle before flowing to the lungs to get re-oxygenated, Choices *A* and *C*.

21. C: Lava is a type of molten rock. When molten rock cools down and hardens, it forms igneous rock. Sedimentary and metamorphic rocks, Choices *A* and *B*, are not formed from molten rock. Choice *D* is incorrect because lava does cool down eventually and becomes hard.

22: A: Looking at Figure 1, the granite is found in column A. The description of the rocks in these columns says that these rocks were formed from molten rock. When molten rock cools, it forms igneous rock. Column C describes how metamorphic rocks are formed, Choice *B*. Column B describes how sedimentary rocks are formed, Choice *C*. Fossils, Choice *D*, are not rocks but are formed into sedimentary rock.

23: D: Sedimentary rocks are formed from soft materials, such as sand, shells, and pebbles. This allows for fossils to form because the remains of animals or plants can be pressed into the softer rock material and leave their imprint. Fossils cannot form in igneous, Choice *A*, or metamorphic, Choice *C*, rocks. Bones are something that can actually make a fossil imprint, Choice *B*.

24. B: Rocks that are formed by changes in heat and pressure are called metamorphic rocks, which is how the rocks in Column C are described. Slate is found in Column C of Figure 1. Sandstone and limestone, Choices *A* and *D*, are both found in Column B, which describes sedimentary rock. Granite, Choice *B*, is found in Column A, which describes sedimentary rock.

25. A: Sedimentary rock is formed from sand, shells, and pebbles, all of which are found in abundance at the beach. Shells, Choice *B*, are something that contribute to the formation of sedimentary rock. Igneous rock, Choice *C*, is formed from molten rock, which would likely be much too hot to be found on most beaches. The surface environment of a beach likely does not undergo changes in heat and pressure enough to form metamorphic rock, Choice *D*.

26. D: Helium gas is not one of the major direct or indirect greenhouse gases. Looking at Figure 2, Choices *A*, *B*, and *C* are part of the pie chart as greenhouse gases found in the atmosphere.

27. B: The two major wedges of the pie chart in Figure 3 are Transportation and Electricity. Using a car, Choice *C*, produces a lot of carbon dioxide. Walking instead of driving a car would not produce any carbon dioxide. Choices *A* and *D* use electricity and leaving either the lights or the television on would need a constant source of electricity, producing lots of carbon dioxide.

28. A: Looking at Figure 2, carbon dioxide, Choice *D*, takes up the largest wedge of the pie chart at 54.7%. The next largest wedge is methane at 30%. Nitrous oxide, Choice *C*, takes up only 4.9% and fluorinated gases, Choice *B*, takes up 0.9%.

29. C: Looking at Figure 1, in the right picture where there are more greenhouse gases, the re-emitted heat arrow is larger as more heat gets trapped in the Earth's atmosphere. Space and the Sun, Choices *A* and *B*, remain outside the Earth's atmosphere. Water vapor, Choice *D*, is not a part of the diagrams for the greenhouse effect.

30. D: Figures 2 and 3 are pie charts. Circular charts that are broken up into wedges, or pie pieces, are called pie charts. Scatter plots, Choice *A*, have specific point markers to mark each data point. In line graphs, Choice *B*, data are represented by connecting lines. In bar graphs, Choice *C*, data are represented by vertical or horizontal bars.

31. A: An anemometer measures air speed. Wind is the movement of air, so an anemometer would be able to measure wind speed. A barometer, Choice *B*, measures air pressure. A thermometer, Choice *C*, measures temperature. A wind vane, Choice *D*, shows which direction the wind is blowing.

32. B: Figure 1 is a graph showing the temperature on different days. Temperature is measured using a thermometer. A rain gauge, Choice *A*, would allow a meteorologist to record amounts of rainfall. A barometer, Choice *C*, measures air pressure. An anemometer, Choice *D*, measures air speed.

33. C: A wind vane shows which direction the wind is blowing. If it is pointing north, the wind is blowing in a northern direction. It would not be blowing in an eastern direction, Choice *A*, or a southern direction, Choice *D*, since that is the opposite direction of north. A wind vane simply tells wind direction and does not determine whether a storm is coming, Choice *B*.

34. A: Looking at the line graph in Figure 2, the lowest point is marked for January, with approximately 2 mm of rainfall. April, Choice *B*, has approximately 23 mm rainfall, September, Choice *C*, has 10 mm of rainfall, and October, Choice *D*, has the highest rainfall at 60 mm.

35. D: Reading the graph in Figure 2, at June, the rainfall is approximately 20 mm. The blue line marks all of the data collected for each month. For June, the blue line is just about at the 20 mm mark from the vertical axis on the left side of the graph.

36. B: Plants use chloroplasts to turn light energy into glucose. Animal cells do not have this ability. Comparing Figures 1 and 2, chloroplasts can be found in the plant cell but not the animal cell.

37. C: The nucleolus is always located inside the nucleus. It contains important hereditary information about the cell that is critical for the reproductive process. Chloroplasts, Choice *A*, are only located in plant cells. It is not found in the mitochondria, Choice *B*, or attached to the cell membrane, Choice *D*.

38. A: Looking at the table in Figure 3, each organelle is described and mitochondria is described as the powerhouse of the cell. The nucleus, Choice *B*, contains the cell's DNA. The ribosomes, Choice *C*, build proteins. The cell wall, Choice *D*, maintains the shape of plant cells and protects its contents.

39. D: Figure 3 describes the functions of the organelles. The cell membrane surrounds the cell and regulates which molecules can move in and out of the cell. Ribosomes build proteins, Choice *A*. Lysosomes, Choice *B*, break down large molecules. The nucleus, Choice *C*, contains the cell's DNA.

40. C: Figure 3 describes the functions of the organelles. Chloroplasts are responsible for photosynthesis in plant cells, which is the process of converting sunlight energy to glucose energy. The cell wall helps maintain the cell's shape, Choice *A*. The nucleus contains the cell's DNA, Choice *B*. Ribosomes build proteins, Choice *D*.

ACT Science Practice Test #2

Passage 1

Questions 1–5 pertain to Passage 1:

Scientists use the scientific method to investigate a theory or solve a problem. It includes four steps: observation, hypothesis, experiment, and conclusion. Observation occurs when the scientist uses one of their senses to identify what they want to study. A hypothesis is a conclusive sentence about what the scientist wants to research. It generally includes an explanation for the observations, can be tested experimentally, and predicts the outcome. The experiment includes the parameters for the testing that will occur. The conclusion will state whether or not the hypothesis was supported.

Scientist A would like to know how sunlight affects the growth of a plant. She says that more sunlight will cause the plant to grow faster. She sets up her experimental groups and tests her hypothesis over 11 days.

Figure 1 below shows the experimental data Scientist A collected over 11 days.

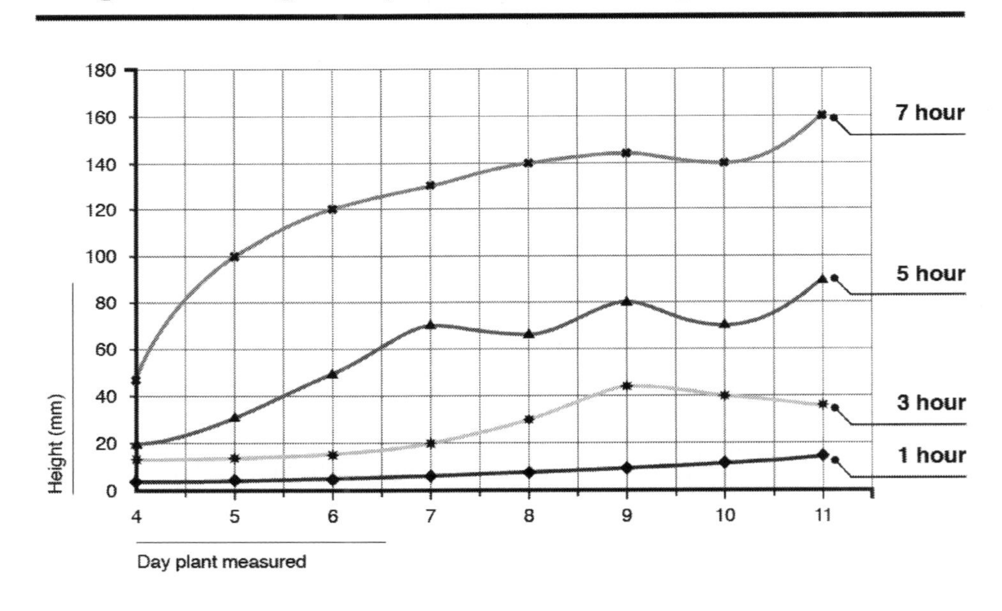

Figure 2 below represents the process of photosynthesis that occurs in plants.

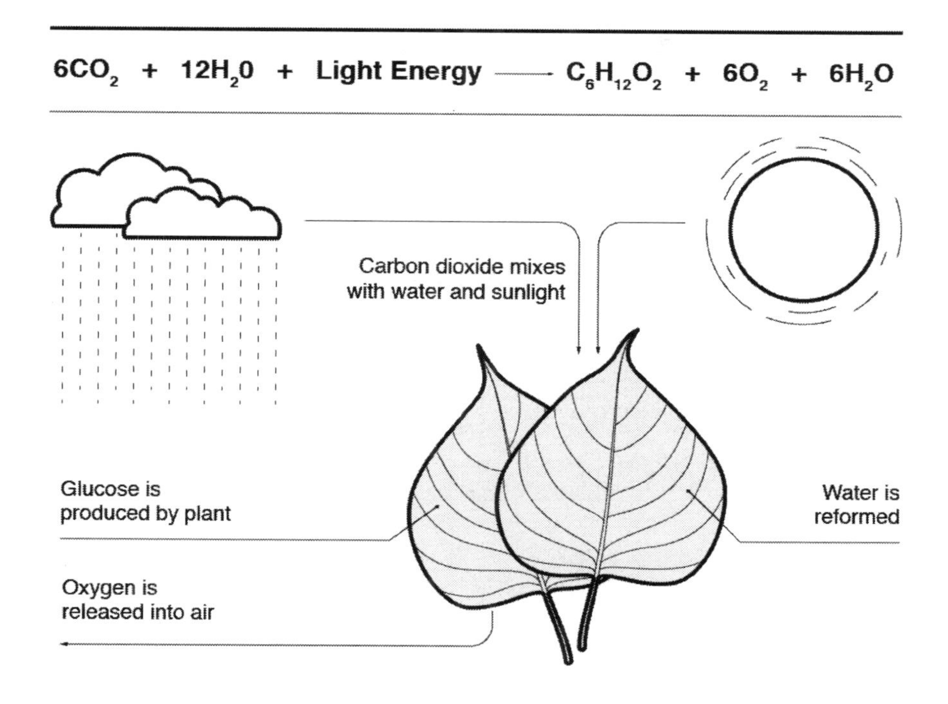

1. What is her hypothesis?
    a. More sunlight will cause the plant to grow faster.
    b. She will test her theory over 11 days.
    c. How sunlight affects plant growth.
    d. Plants do not grow well with one hour of sunlight per day.

2. How many experimental groups does she have?
    a. 1
    b. 3
    c. 4
    d. 11

3. What type of chart is represented in the first figure?
    a. Bar graph
    b. Line graph
    c. Pie chart
    d. Pictogram

4. What part of the photosynthesis reaction is provided directly by sunlight?
    a. Light energy
    b. H_2O
    c. CO_2
    d. Glucose

5. What should her conclusion be based on her experimental data?
    a. 5 hours of sunlight is optimal for plant growth.
    b. Plants should only be measured for 11 days.
    c. Less sunlight is better for plant growth.
    d. Providing plants with more sunlight makes them grow bigger.

Passage 2

Questions 6–10 pertain to Passage 2:

The periodic table contains all known 118 chemical elements. The first 98 elements are found naturally while the remaining were synthesized by scientists. The elements are ordered according to the number of protons they contain, also known as their atomic number. For example, hydrogen has an atomic number of one and is found in the top left corner of the periodic table, whereas radon has an atomic number of 86 and is found closer on the right side of the periodic table, several rows down. The rows are called periods and the columns are called groups. The elements are arranged by similar chemical properties.

Each chemical element represents an individual atom. When atoms are linked together, they form molecules. The smallest molecule contains just two atoms, but molecules can also be very large and contain hundreds of atoms. In order to find the mass of a molecule, the atomic mass of each individual atom in the molecule must be added together.

Figure 1 below depicts the trends and commonalities between the elements that can be seen in the periodic table.

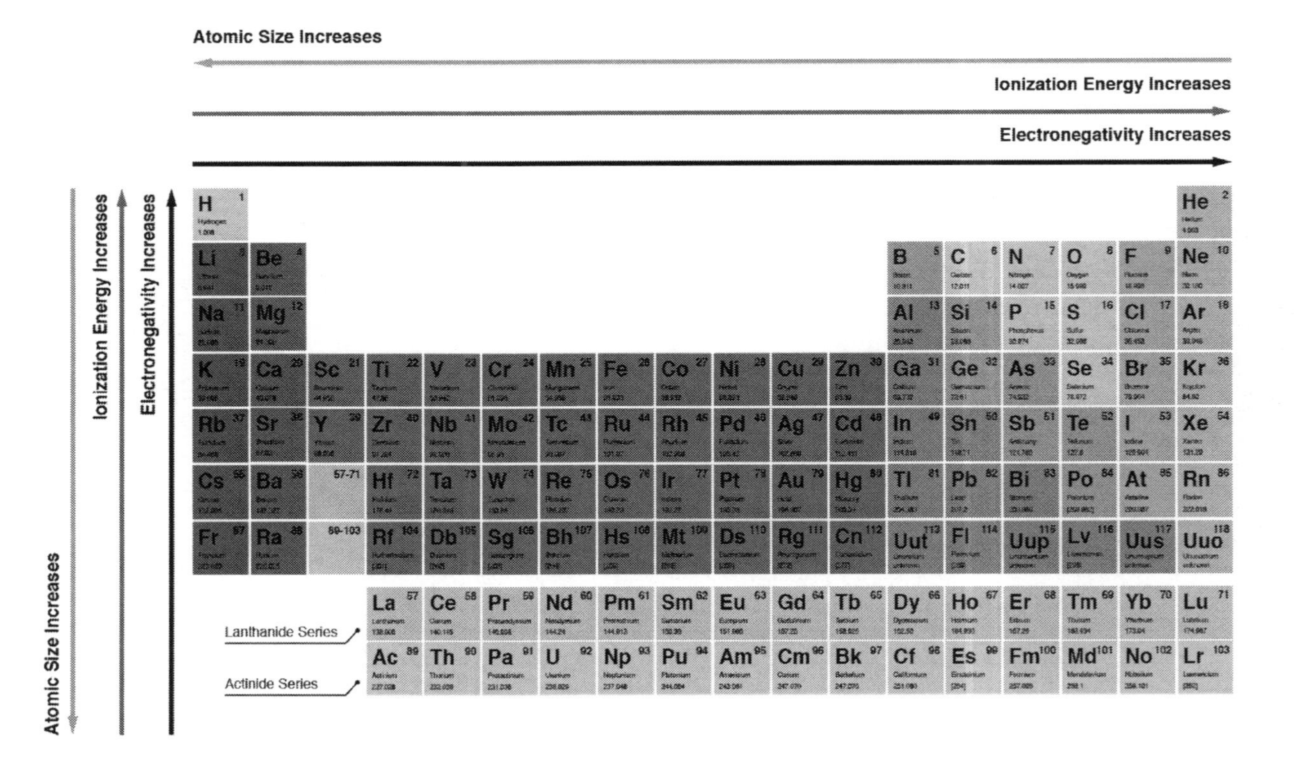

Figure 2 below shows what the information in each element's box represents.

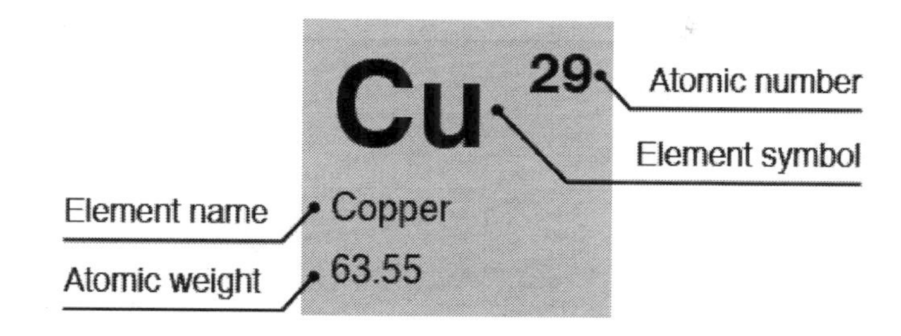

Figure 3 below shows the periodic table with color coding according to the groups and periods.

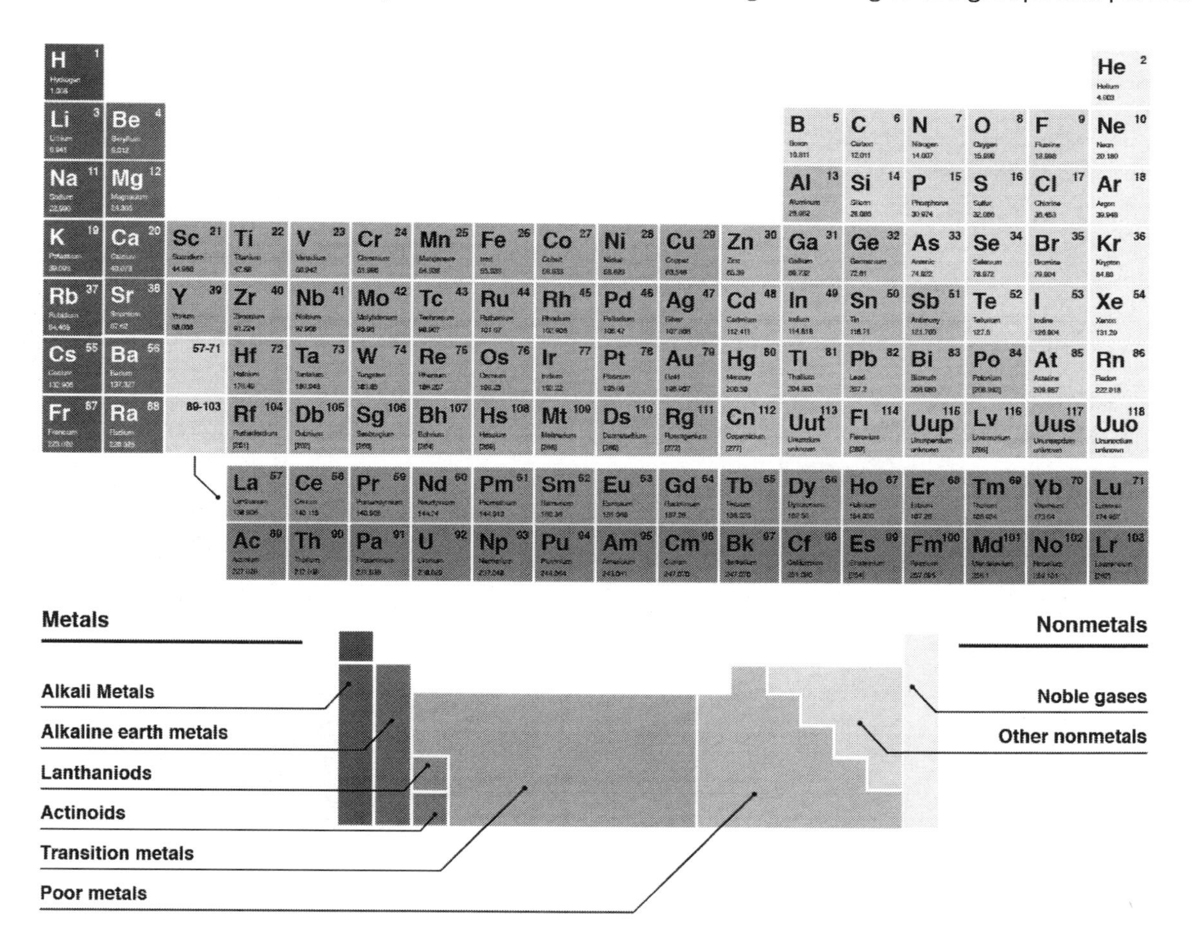

6. What is the atomic mass of NaCl?
   a. 23
   b. 58.5
   c. 35.5
   d. 71

7. Which of the following elements is most electronegative?
   a. Ununoctium (Uuo)
   b. Francium (Fr)
   c. Hydrogen (H)
   d. Helium (He)

8. What is the full name of the element Cr?
   a. Chromium
   b. Copper
   c. Chlorine
   d. Curium

9. Which element has the fewest number of protons?
   a. Radon (Rn)
   b. Boron (B)
   c. Nitrogen (N)
   d. Hydrogen (H)

10. Scientist A needs a noble gas for her experiment. Which of these elements should she consider using?
   a. Nitrogen (N)
   b. Radon (Rn)
   c. Copper (Cu)
   d. Boron (B)

Passage 3

Questions 11–15 pertain to Passage 3:

> Physical characteristics are controlled by genes. Each gene has two alleles, or variations. Generally, one allele is more dominant than the other allele and when one of each allele is present on the gene, the physical trait of the dominant allele will be expressed. The allele that is not expressed is called the recessive allele. Recessive alleles are expressed only when both alleles present on the gene are the recessive allele.
>
> Punnett squares are diagrams that can predict the outcome of crossing different traits. In these diagrams, dominant alleles are represented by uppercase letters and recessive alleles are represented by lowercase letters.
>
> Scientist A wants to grow white flowered plants and is doing a series of crossbreeding experiments. She had each plant genetically tested so she knows which alleles comprise each plant. The dominant flowers are red (A) and the recessive allele (a) produces white flowers.

Figure 1 below represents the different flowers that underwent crossbreeding during Round #1A

Round #1A

Crossbreeding #1A			Crossbreeding #2A			Crossbreeding #2A		
	A	**a**		**a**	**a**		**a**	**a**
A	AA	Aa	**A**	Aa	Aa	**A**	Aa	Aa
A	AA	Aa	**A**	Aa	Aa	**a**	aa	aa

Figure 2 below represents the number of flowers that were red and white after the first round of crossbreeding experiments.

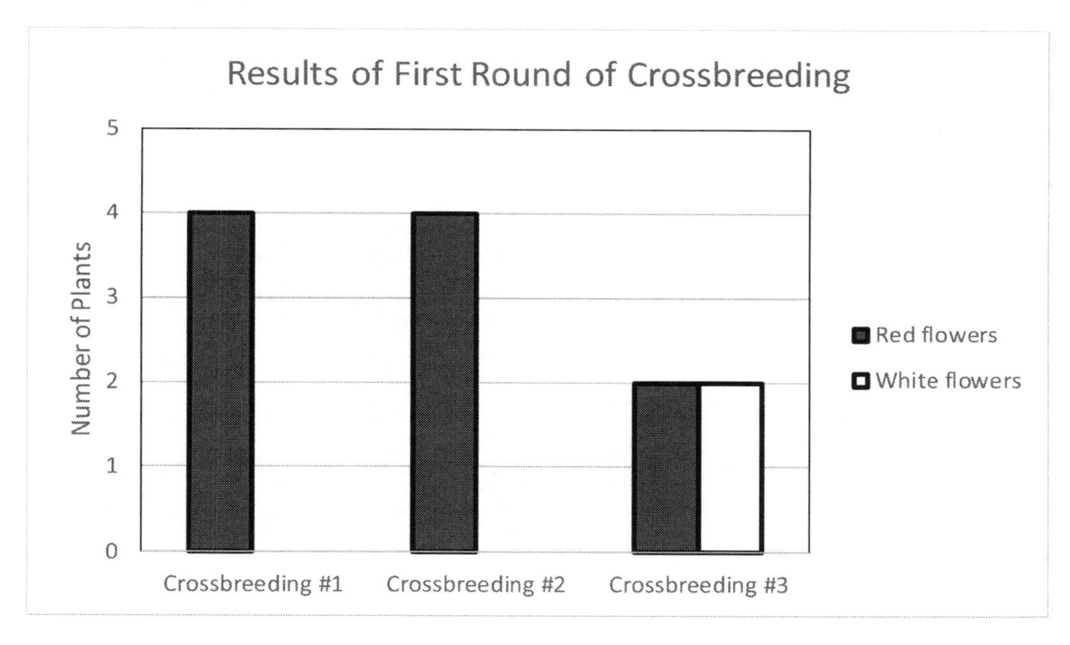

During her second round of crossbreeding, she adds in a plant of unknown genetic makeup with red flowers. She crosses it with a white-flowering plant. The results of her experiment are represented in the next figure.

Figure 3 represents the genetic results from the second round of crossbreeding.

Round #2

	a	a
?	Aa	Aa
?	aa	aa

Scientist A takes offspring plants from Round #1A and crossbreeds them with each other and calls this Round #1B.

Figure 4 below represents the results of crossbreeding from Round #1B.

Round #1B

Crossbreeding #1B

	A	a
A	AA	Aa
A	AA	Aa

Crossbreeding #2B

	A	a
A	AA	Aa
a	Aa	aa

Crossbreeding #2B

	a	a
a	aa	aa
a	aa	aa

11. Crossbreeding which two plants will give her the highest likelihood of obtaining some white plants right away in Round #1A?
    a. AA × Aa
    b. Aa × aa
    c. AA × aa
    d. AA × AA

12. What percentage of plants are white after the first crossbreeding reactions?
    a. 12
    b. 50
    c. 16.7
    d. 25

13. What is the genetic makeup of the unknown plant from the second round of crossbreeding?
    a. aa
    b. Aa
    c. AA
    d. Cannot be determined

14. From which group of crossbreeding in Round #1B can she obtain 100% white flowers by the second generation?
   a. They are all equal.
   b. 1B
   c. 2B
   d. 3B

15. Which of her five senses did she use for the observation step of the scientific method here?
   a. Sight
   b. Smell
   c. Touch
   d. Hearing

Passage 4

Questions 16–20 pertain to Passage 4:

> Rainforests cover approximately 6% of the Earth's surface. Tropical rainforests are found in five major areas of the world: Central America, South America, Central Africa, Asia stretching from India to islands in the Pacific Ocean, and Australia. All of these areas are warm and wet areas within ten degrees of the equator. They do not have a substantial dry season during the year.
>
> Rainforests are large areas of jungle that get an abundance of rain. They comprise four layers, each with unique characteristics. The emergent layer is the highest layer and is made up of the tops of the tall trees. There is very good sunlight in this layer. The canopy layer is the next layer, just under the emergent layer. Here, there is some sun but not as much as the emergent layer. The next layer is the understory layer. This layer does not receive very much sunlight. The plants in this layer need to grow very large leaves to reach the sun. The bottom-most layer is the forest floor. Sunlight generally does not reach this layer, so plants do not grow here.

Figure 1 below represents the different layers of the rainforest.

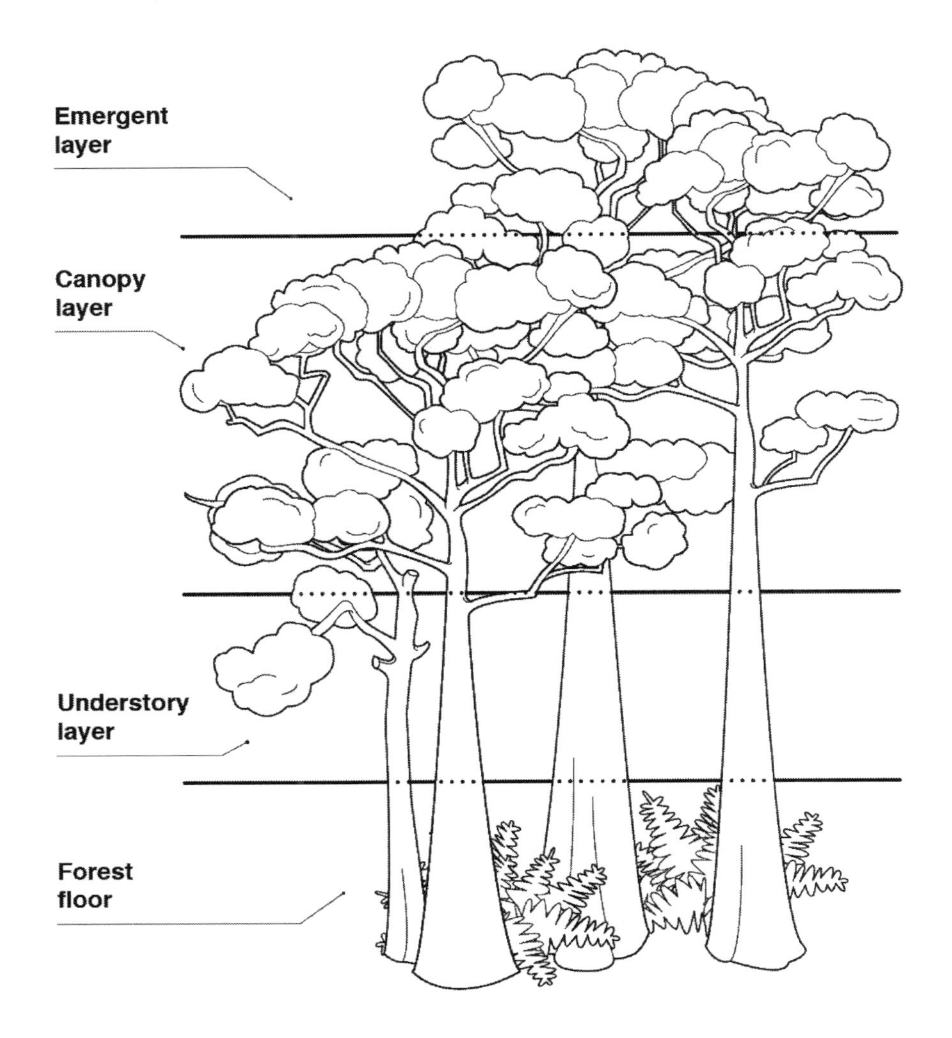

Figure 2 below is a map of the rainforests on Earth and a map of Central America.

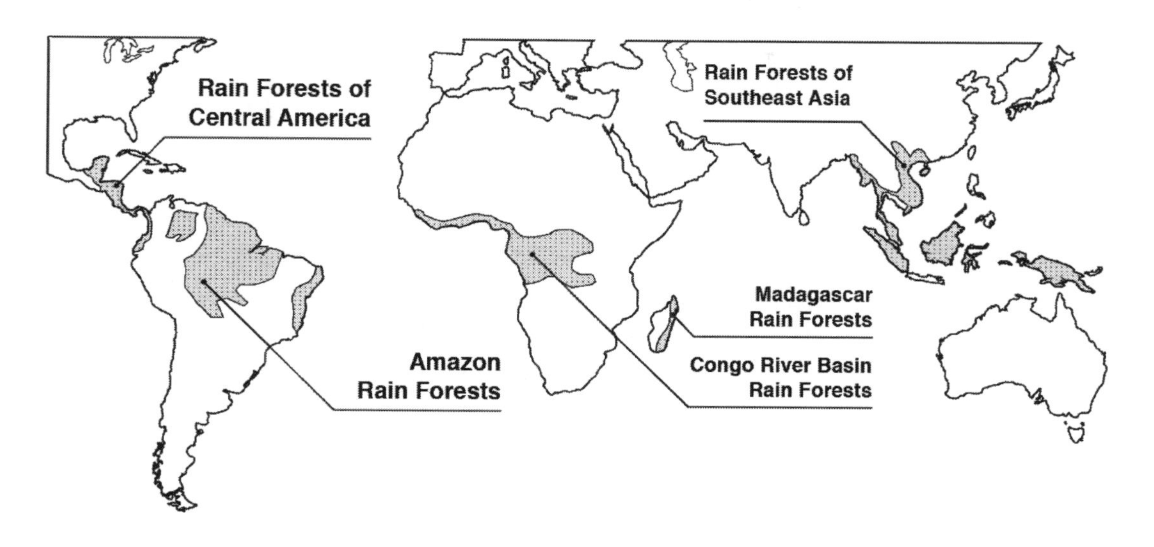

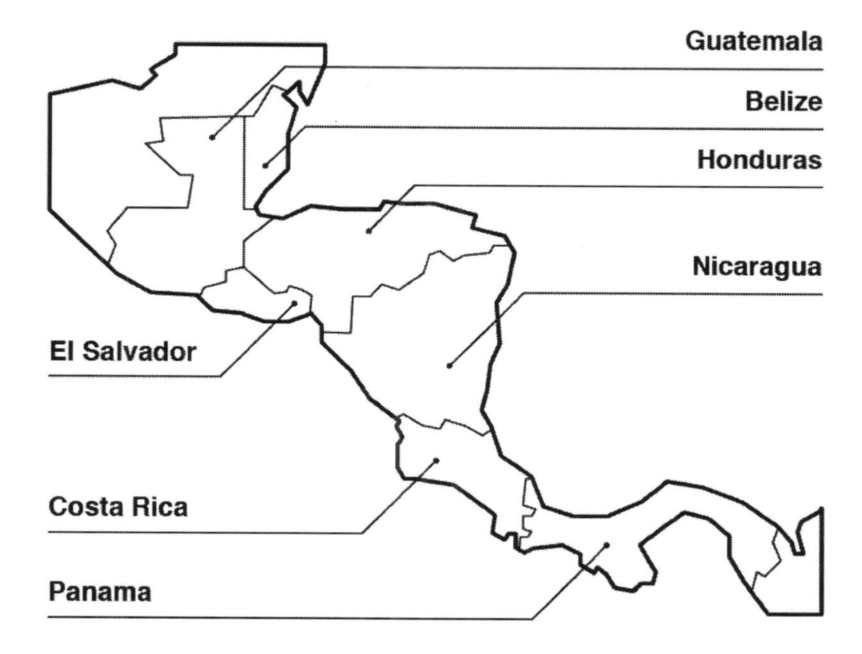

16. What essential part of photosynthesis, which is necessary for plant growth, is lacking on the forest floor and does not allow for plants to grow?

a. Carbon
b. Water
c. Oxygen
d. Sunlight

17. Which is a country in Central America that contains a rainforest?
    a. India
    b. Panama
    c. Madagascar
    d. Brazil

18. In which layer of the rainforest would birds fly around the most?
    a. Forest floor
    b. Understory layer
    c. Emergent layer
    d. Canopy layer

19. Giant taro plants have the largest leaves in the world that are approximately ten feet in length. In which layer of the rainforest do they reside?
    a. Emergent layer
    b. Canopy layer
    c. Forest floor
    d. Understory layer

20. Which of these plants, based on their listed requirements, would thrive in a rainforest climate?
    a. Giant water lily: warm temperatures, wet environment, has ability to grow large leaves
    b. Cactus: dry environment, long, hot season for growth
    c. Pine tree: dry and sandy soil, lots of sunlight
    d. Black-eyed sun: very hot temperatures, slightly moist soil

Passage 5

Questions 21–25 pertain to Passage 5:

An eclipse occurs when the light from one object in the solar system is completely or partially blocked by another object in the solar system. In 2017, the Earth was a part of two different types of eclipses. One was a solar eclipse, which occurs when the moon passes between the Sun and the Earth and blocks the Sun's light, making it dark during the daytime for several minutes. A total solar eclipse occurs when the moon completely covers the Sun. A partial solar eclipse occurs when the moon only covers part of the Sun. Solar eclipses should not be looked at directly because the Sun's rays can damage a person's eyes even though they appear to be dim while the eclipse is happening. Special viewing devices can be used to look at a solar eclipse indirectly, such as a pinhole camera facing away from the eclipse that allows light from the eclipse to pass through a hole in a piece of cardboard and its image to be reflected on a piece of white paper.

The other type of eclipse that the Earth was a part of was a lunar eclipse. This type of eclipse occurs when the moon passes behind the Earth, into its shadow. The moon is illuminated by the light of the sun, so when it is in the Earth's shadow, the moon becomes dim for a few hours during the night. This type of eclipse is safe to look at without any protection for your eyes.

Figure 1 below represents a solar eclipse.

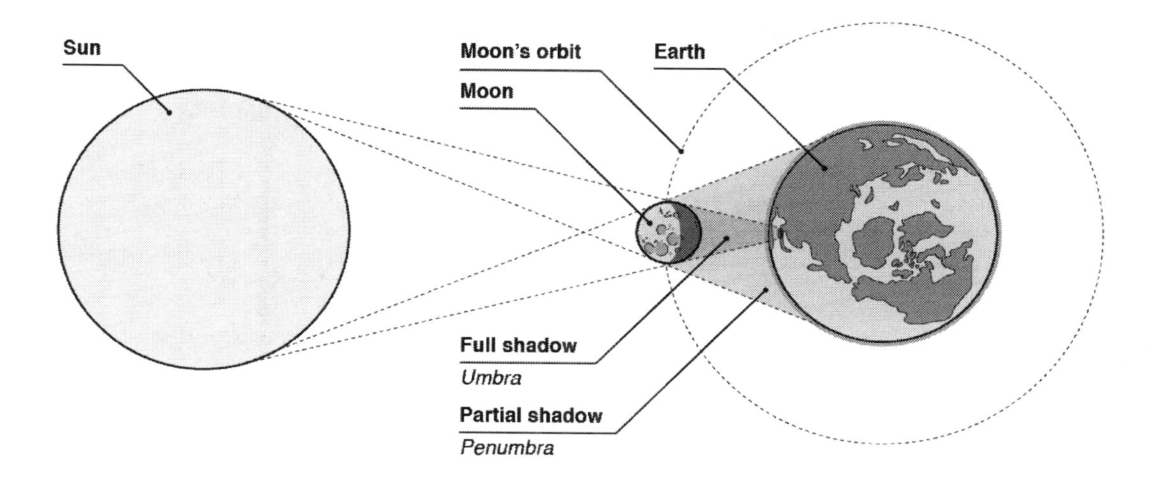

Figure 2 below represents a lunar eclipse.

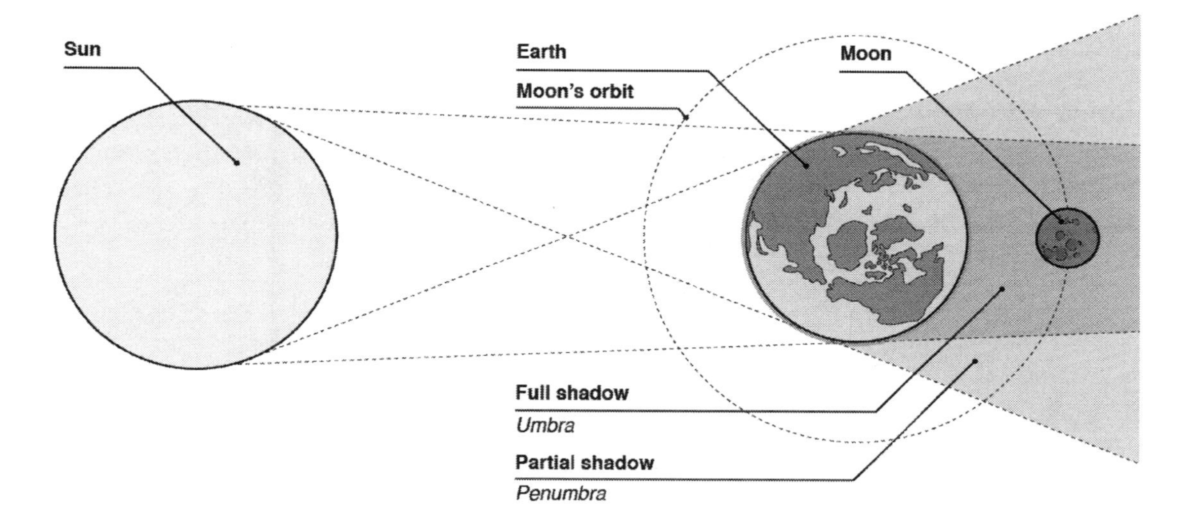

21. Which apparatus would be best to use to look at a solar eclipse?
    a. A telescope facing the eclipse
    b. A pinhole camera facing away from the eclipse
    c. Sunglasses facing the eclipse
    d. Binoculars facing the eclipse

22. What type of eclipse occurs when the moon comes between the Earth and Sun and covers the Sun's light completely?
    a. Total solar eclipse
    b. Partial lunar eclipse
    c. Total lunar eclipse
    d. Partial solar eclipse

23. What object in the solar system becomes dim during a lunar eclipse?
    a. Sun
    b. Earth
    c. Moon
    d. Earth and moon

24. Which type of eclipse could you observe directly using a telescope?
    a. Neither solar nor lunar
    b. Lunar only
    c. Both solar and lunar
    d. Solar only

25. Which type of eclipse is viewed during the daytime?
    a. Both solar and lunar
    b. Solar only
    c. Partial lunar
    d. Total lunar

Passage 6

Questions 26–30 pertain to Passage 6:

> Phylogenetic trees are diagrams that map out the proposed evolutionary history of a species. They are branching diagrams that make it easy to see how scientists believe certain species developed from other species. The most recent proposed common ancestor between two species is the one before their lineages branch in the diagram. These diagrams do not attempt to include specific information about physical traits that were thought to be retained or disappeared during the evolutionary process.
>
> Cladograms classify organisms based on their proposed common ancestry but are focused on their common physical traits. Branching points on these diagrams represent when a group of organisms is thought to have developed a new trait. Analogous features are those that have the same function but were not derived from a common ancestor. Homologous features have anatomical similarities, even if the function is no longer the same, due to a proposed common ancestor.

Figure 1 below is a phylogenetic tree of the Carnivora order.

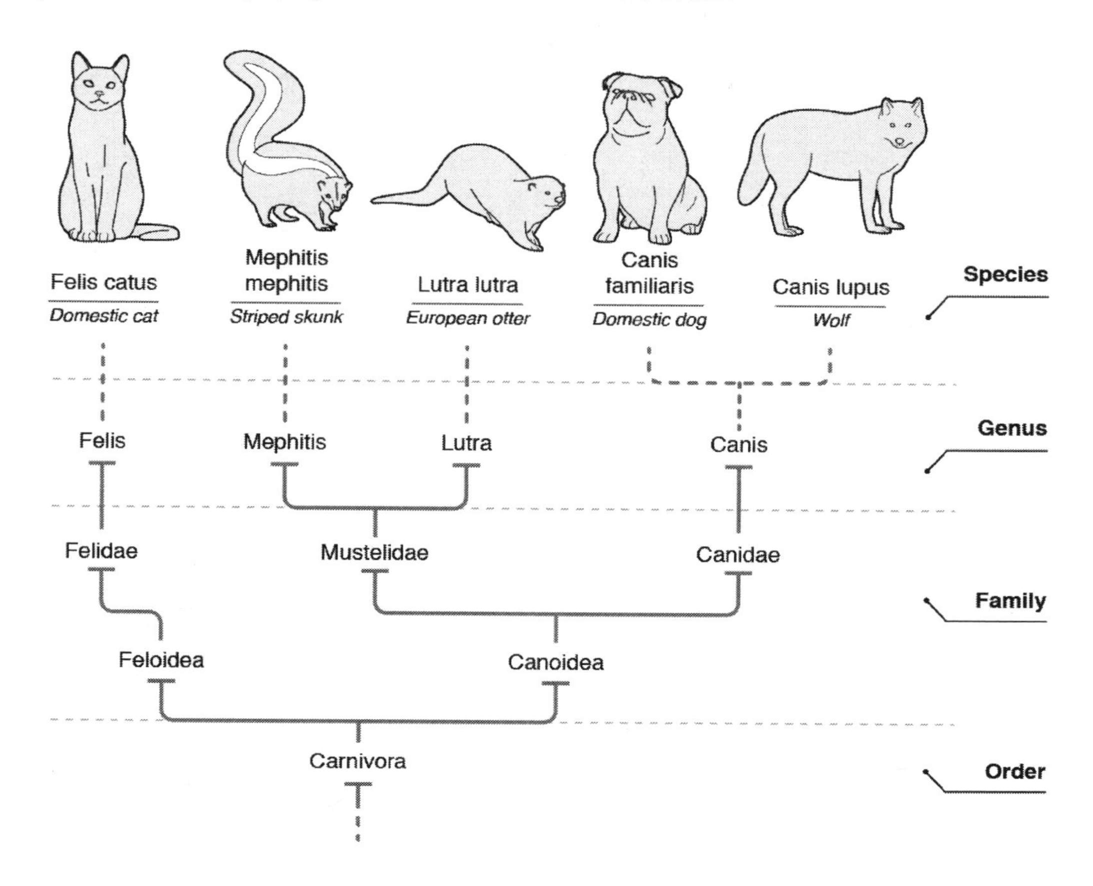

Figure 2 below is a cladogram.

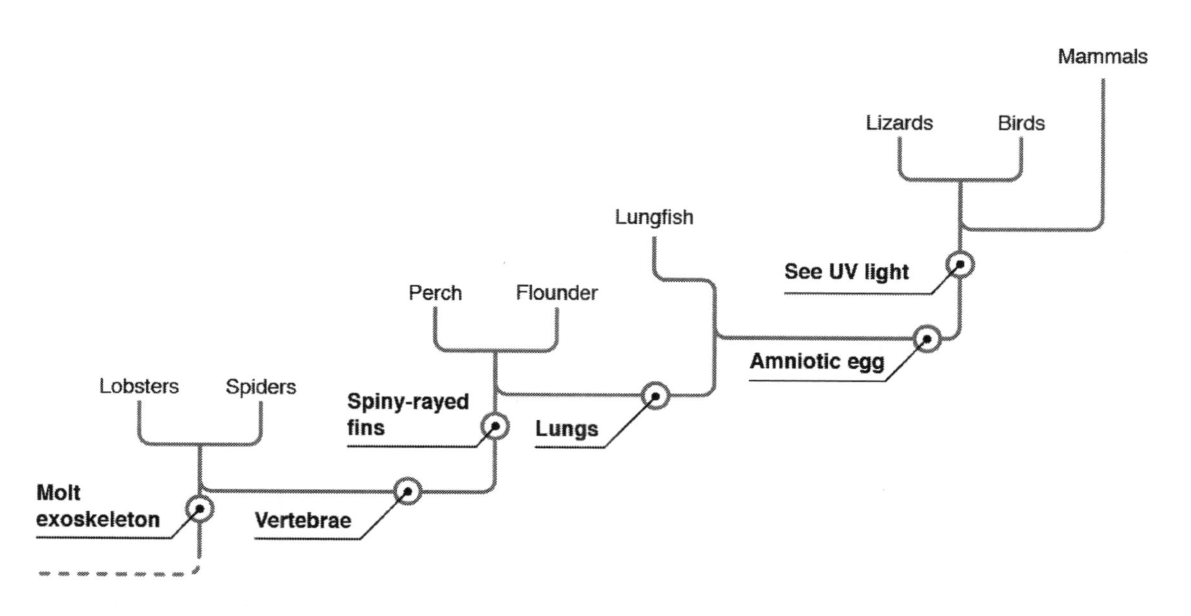

Figure 3 below shows a homologous feature between four different species with a proposed common ancestor.

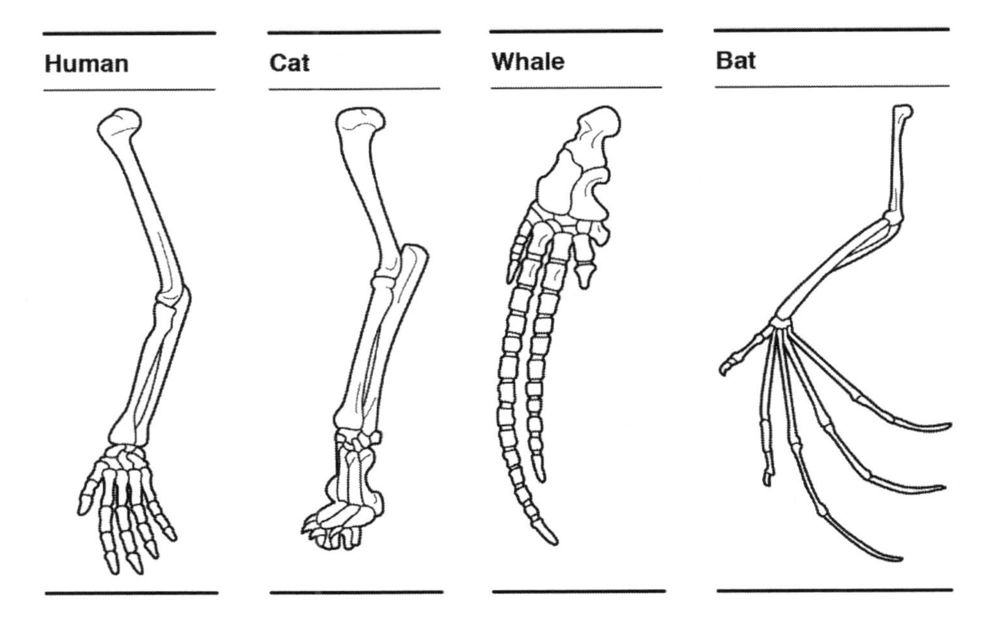

26. Which tool would you use to find out when a common ancestor of two species supposedly developed lungs?
    a. Phylogenetic tree
    b. Cladogram
    c. Punnett square
    d. Photographs

27. According to the earlier details, how are human arms and whale fins related?
    a. They are homologous structures from a common ancestor.
    b. They both have the same number of bones.
    c. They are analogous features.
    d. They are both covered in the same type of skin.

28. According to Figure 1, what common ancestry group do the striped skunk and European otter share?
    a. Mephitis
    b. Felidae
    c. Canidae
    d. Mustelidae

29. What trait do lizards and birds have in common according to Figure 2?
    a. Both see UV light
    b. Both have spiny-rayed fins
    c. Both molt an exoskeleton
    d. They do not have any traits in common.

30. According to Figure 1, at what level of organization are domestic cats and wolves related?
    a. Family
    b. Genus
    c. Order
    d. Species

Passage 7

Questions 31–35 pertain to Passage 7:

Scientists often use an assay called an enzyme-linked immunosorbent assay, or ELISA, to quantify specific substances within a larger sample. An ELISA works based on the specificity of an antibody to an antigen. One type of ELISA is called a sandwich ELISA. In this type of ELISA, a plate is coated with a capture antibody that adheres the antigen in the sample when it is added. Then the primary antibody is added and sticks to any antigen bound to the capture antibody. Next, a secondary antibody is added. Once it attaches to the primary antibody, it releases a colored tag that can be detected by a piece of laboratory equipment. If more color is released, it is indicative of more antigen having been present in the sample.

Figure 1 below describes how a sandwich ELISA works.

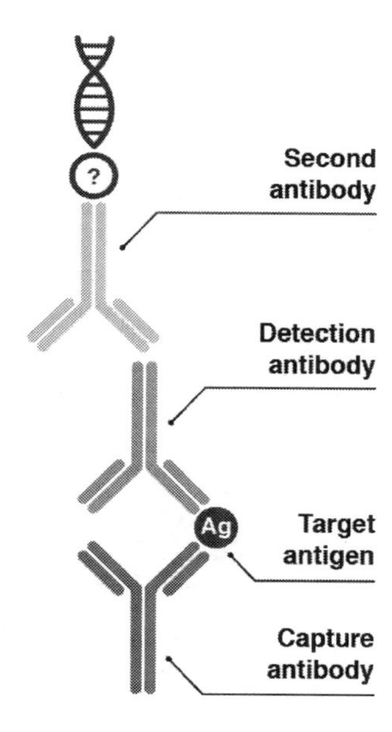

The cytokine protein IL-1β is a marker of inflammation in the body. Scientist A took samples from different locations within the body to find out where there was elevated inflammation in a patient.

Figure 2 below is a picture of the ELISA plate from Scientist A's experiment.

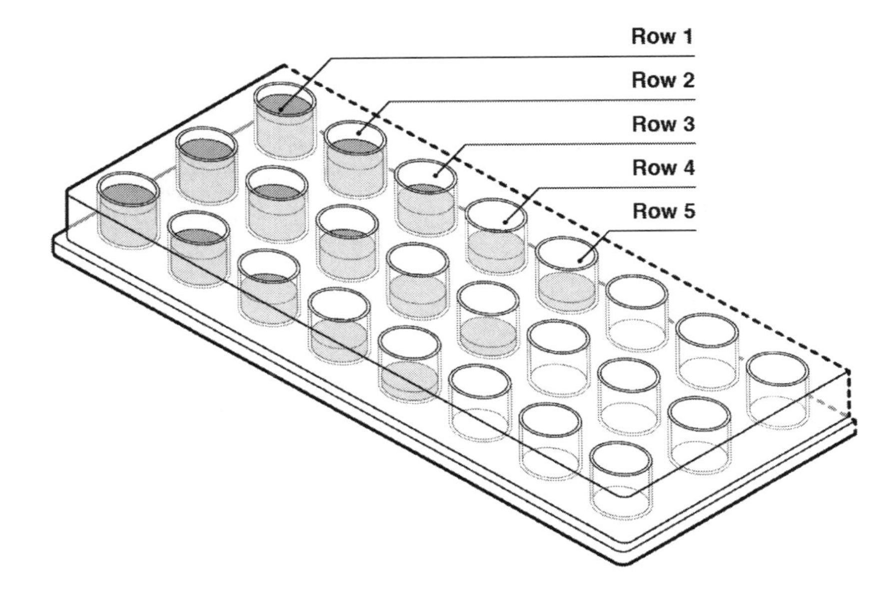

Figure 3 below is a graph of the results of Scientist A's experiment.

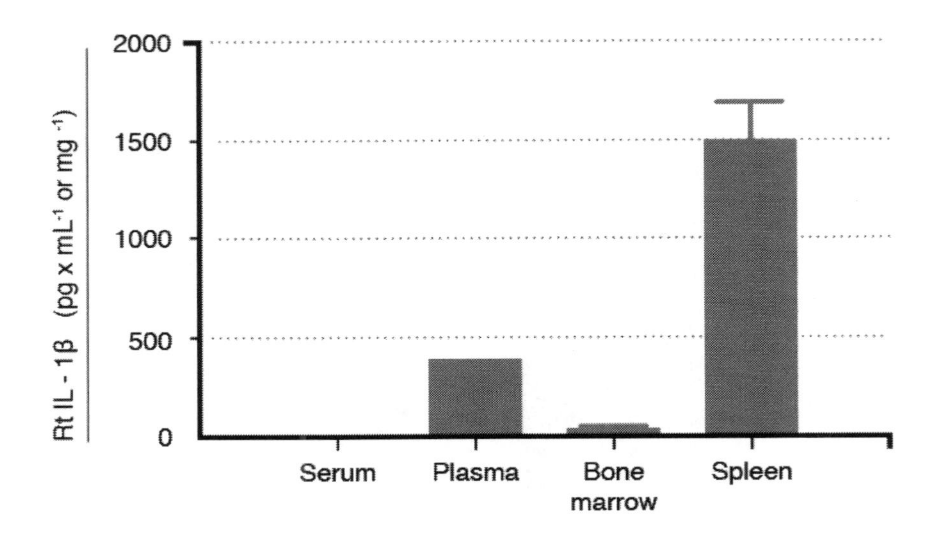

31. Which step of the ELISA allows for the color to be released for detection of the antigen?
    a. Addition of the antigen
    b. Addition of the primary antibody
    c. The presence of the capture antibody
    d. Addition of the secondary antibody

32. According to Figure 2, which row had the largest amount of antigen in the sample?
    a. Row 2
    b. Row 1
    c. Row 5
    d. Row 3

33. According to the ELISA results in Figure 3, which area of the body had the most inflammation?
    a. Serum
    b. Plasma
    c. Spleen
    d. Bone marrow

34. Which two antibodies sandwich the antigen in a sandwich ELISA?
    a. Capture antibody and secondary antibody
    b. Capture antibody and primary antibody
    c. Primary antibody and secondary antibody
    d. Two units of the secondary antibody

35. What is the purpose of an ELISA?
    a. Quantify specific substances within a larger sample
    b. Quantify all substances within a larger sample
    c. To create a colorful pattern with the samples
    d. To develop different antibodies

Passage 8

Questions 36–40 pertain to Passage 8:

> Natural selection is the idea that certain traits make an individual have longer survival and higher reproduction rates than other individuals. It is based on the **phenotype**, or physical appearance, of the individual and not the **genotype**, or genetic makeup. There are three ways in which a phenotype can change due to natural selection. Directional selection occurs when one extreme of a phenotype is favored. Disruptive selection occurs when both extremes of a phenotype are favored. Stabilizing selection occurs when an intermediate phenotype is favored over either extreme phenotype.
>
> *Scenario 1:* Mice live in an environment that has a mix of light and dark colored rocks. To avoid predators, the mice with intermediate color fur survive longer and produce more offspring.
>
> *Scenario 2:* The Galapagos Islands experienced a drought and large, tough seeds became abundant. Finches developed large beaks to break up these seeds.
>
> *Scenario 3:* In Cameroon, seeds are either large or small. Finches in Cameroon have either large beaks or small beaks. They are not found with medium-sized beaks.

36. What type of selection is described in Scenario 1?
    a. Stabilizing selection
    b. Directional selection
    c. Disruptive selection
    d. Color selection

37. What type of selection is described in Scenario 2?
    a. Beak-type
    b. Disruptive
    c. Stabilizing
    d. Directional

38. Why would it be hard for small-beaked finches in the Galapagos Island to survive after the drought?
    a. Too much sand got caught in the small beaks
    b. Beaks would not be able to break up large seeds
    c. Large-beaked finches would attack them
    d. Two extreme phenotypes can never be selected by natural selection

39. What type of selection is described in Scenario 3?
    a. Stabilizing
    b. Directional
    c. Disruptive
    d. Beak-type

40. Which statement is true about natural selection?
    a. Individuals are selected based on their genotype.
    b. An extreme phenotype is always selected.
    c. It only occurs after a drought.
    d. Individuals are selected based on phenotypes that are advantageous for survival and reproduction.

Answer Explanations #2

1. A: The hypothesis is the sentence that describes what the scientist wants to research with a conclusive expected finding. Choice *A* describes how she believes sunlight will affect plant growth. Choice *B* includes details about the experiment. Choice *C* is not a conclusive theory. Choice *D* describes the data that she found after conducting the experiment.

2. C: Looking at Figure 1, four experimental groups are shown on the graph for which data were collected: plants that received 1 hour of sunlight, 3 hours of sunlight, 5 hours of sunlight, and 7 hours of sunlight. Choices *A* and *B* could be describing two of the experimental groups and how much sunlight they received. Choice *D* describes how many days' data was collected.

3. B: After the data was collected, it was compiled into a line graph. The data points were collected, and then a line was drawn between the points. Data is represented by horizontal or vertical bars in bar graphs, Choice *A*. Pie charts are circular charts, with the data being represented by different wedges of the circle, Choice *C*. Pictograms use pictures to describe their subject, Choice *D*.

4. A: Looking at Figure 2, the sun provides light energy that drives forward the process of photosynthesis, which is how plants make their own source of energy and nutrients. Choices *B* and *C* are found in the environment around the plants. They combine with light energy to make the photosynthesis reaction work. Choice *D* is a product of photosynthesis.

5. D: Looking at the Figure 1, the experimental group that received 7 hours of sunlight every day grew taller than any of the other groups that received less sunlight per day. Therefore, it is reasonable to conclude that more sunlight makes plants grow bigger. Choice *A* is not a reasonable conclusion because it did not have the tallest plants. The scientist decided to measure the plants only for 11 days, but that does not describe a conclusion for the experiment, Choice *B*. Choice *C* is the opposite of the correct conclusion and does not have evidence to support it.

6. B: The atomic mass of a molecule can be found by adding the atomic mass of each component together. Looking at Figure 2, the atomic mass of each element is found below its symbol. The atomic mass of Na is 23, Choice *A*, and the atomic mass of Cl is 35.5, Choice *C*. The sum of those two components is 58.5, Choice *B*. Choice *D* is equal to two Cl atoms joined together.

7. D: Figure 1 shows the trends of the periodic table. Looking at the black arrows representing electronegativity, it is shown that electronegativity increases going towards the top row of the table and also increases going towards the right columns of the table. Therefore, the most electronegative element would be found in the top right corner of the table, which is where the element Helium is found. Choices *A* and *B* are found at the bottom of the table. Choice *D* is found on the left side of the table.

8. A: Looking at Figure 2, the element name is found under the symbol in each box on the periodic table. Looking at Figure 1 or 3, the full name of element Cr is Chromium. Copper, Choice *B*, is represented by Cu. Chlorine, Choice *C*, is represented by Cl. Curium, Choice *D*, is represented by Cm.

9. D: The atomic number of an element represents the number of protons. Looking at Figure 2, the atomic number is located at the top of the box, above the element's symbol. Hydrogen (H) has an atomic number of 1 and has the least number of protons of any other element in the periodic table.

Radon (Rn), Choice *A*, has 86 protons. Boron (B), Choice *B*, has 5 protons. Nitrogen (N), Choice *C*, has 7 protons.

10. B: Looking at Figure 3, the elements are color coded in periods and groups according to their similar properties. Noble gases are located in the right most column of the table. Radon (Rn) is the only one of the element choices marked as a noble gas and would be the right choice for Scientist A. Nitrogen (N) and Boron (B), Choices *A* and *D*, are nonmetals. Copper (Cu), Choice *C*, is a transition metal.

11. B: Looking at Figures 1 and 2, crossbreeding experiment #3 in round #1 produces plants that are completely recessive and would have white flowers. Choices *A* and *C*, crossbreeding experiments #1 and 2, respectively, only produce flowers with a dominant allele present, making red flowers. Choice *D* does not have any recessive alleles, so white flowers are not a possibility.

12. C: Looking at Figure 2, which represents the number of plants that were produced from each crossbreeding experiment, it can be seen that only 2 plants produced white flowers out of 12 plants total, 4 from each experiment. To find the percentage, divide 2 by 12 and multiply by 100. The result is 16.7%. Choice *A* is the total number of plants that were produced. Choice *B* represents the percentage of white flowers in experiment #3 alone.

13. B: In a Punnett Square, each box represents one allele from each of the parent's genes. To find the genetic makeup of the second parent, take out the allele that was contributed from the first parent. Here, the first parent contributed a recessive allele, a, to each offspring. In the top row, that leaves a dominant allele, A, and in the bottom row, that leaves a recessive allele, a. Therefore, the genetic makeup of the second parent is Aa.

14. D: Crossbreeding the plants with only recessive alleles will result in 100% white flowering plants. All four offspring have white flowering plants. Choice *B* gives 100% red flowering plants. Choice *C* gives 25%, 1 out of 4 plants, with white flowers.

15. A: The observation step of the scientific method involves using your senses to identify the results of the experiment. In this case, the experiment depended on identifying the color of the flowers. This was done using sight. If the experiment had involved different scents produced by the flowers, Choice *B* would have worked. If it has involved different textures of the flowers, Choice *C* would have worked. Flowers generally do not make any noise, so Choice *D* would not have been useful.

16. D: The process of photosynthesis requires carbon dioxide and water to combine with sunlight to produce glucose, which is used as an energy source by plants. The forest floor does not get a lot of sunlight since it is shaded by the growth of so many trees and plants in the rainforest. Carbon, Choice *A*, is available through the air. Plants expel carbon dioxide. Water, Choice *B*, is abundant in the humid climate of the rainforest. Oxygen, Choice *C*, is always available in the Earth's atmosphere.

17. B: Central America is one of the five major areas of the world that has a rainforest. Looking at Figure 2, it can be seen that the southern countries of Central America contain rainforests. Comparing this map to the map of Central America, it is clear that Panama is a country that has rainforests. Choices *A*, *C*, and *D* are not found in Central America.

18. C: Birds fly above the trees of the rainforest the most. There, they have unobstructed skies, unlike the dense growth of the trees and plants in the other layers of the rainforest.

19. D: The understory layer is the third layer from the top of the rainforest. It does not receive much sunlight, so the plants need to grow large leaves to absorb as much sunlight as possible. Giant taro leaves would grow well in this layer since they have large leaves. The emergent layer, Choice *A*, gets plenty of sunlight since it is the topmost layer. The canopy layer, Choice *B*, receives enough sunlight for plants to grow without needing to increase their leaf size. The forest floor, Choice *C*, does not receive sunlight, and plants generally do not grow here.

20. A: Rainforests have warm and wet climates. They do not have long dry seasons and tend to have temperate temperatures. The giant water lily is ideal for the rainforest because it can grow large leaves and needs a wet environment to grow in. They grow in the shallow basins of rainforest rivers. Choices *B* and *C* need dry environments. Choice *D* needs a very hot environment, which is not characteristic of rainforests.

21. B: Solar eclipses should not be looked at directly. The rays of the Sun do not seem as bright as normal but can still cause damage to the eyes. A pinhole camera facing away from the eclipse allows the viewer to see a reflection of the eclipse instead of the actual eclipse. Choices *A, C,* and *D* all require looking directly at the solar eclipse.

22. A: When the moon comes between the Earth and Sun, a solar eclipse occurs. If the sun is far enough away and is completely blocked by the moon, it is a total solar eclipse. If it is only partially blocked by the moon, it is a partial solar eclipse, Choice *D*. A lunar eclipse occurs when the moon is on the opposite side of the Earth as the Sun and the Sun creates a shadow of the Earth on the moon, so that the moon becomes completely dark, Choice *C*, or partially dark, Choice *B*.

23. C: During a lunar eclipse, the Sun and moon are on opposite sides of the Earth. They line up so that the Sun's light that normally illuminates the moon is blocked by the Earth. This causes the moon to become dim. Sunlight can still be seen, Choice *A*, and the Earth does not become dark, Choices *B* and *D*.

24. B: The moon does not produce harmful light rays that can damage the eyes, so lunar eclipses can be viewed directly. A telescope would allow the lunar eclipse to be magnified and seen more clearly. During a solar eclipse, the Sun's rays appear to be dim and easy to see directly but they are still harmful to the eyes.

25. B: Solar eclipses are viewed during the daytime because they involve viewing the Sun while it is out during normal daytime hours. Lunar eclipses, Choices *C* and *D*, are viewed at nighttime when the moon is in the sky during its normal hours. The moon is normally illuminated by the Sun that is on the other side of the Earth. When the Sun is on the other side of the earth, it is nighttime for people looking at the moon.

26. B: A cladogram is a diagram that organizes proposed ancestral relations based on the development of physical features. A branching point would be seen on the cladogram where the development of lungs was noted. A phylogenetic tree, Choice *A*, does not note phenotypic features on it. Punnett squares, Choice *C*, are used to determine the possible genetic makeup of offspring and are not related to evolution. Photographs, Choice *D*, may reveal species that look alike but would not reveal if they truly had a common ancestor.

27. A: According to Figure 3, human arms and whale fins are homologous structures that were derived from a common ancestor. They have anatomical similarities, although their function is not the same. They have different numbers of bones, so Choice *B* is incorrect. Since they are proposed to be developed from a common ancestor, they are not analogous features, Choice *C*. Whales have blubber covering their bodies and not layered skin like humans, so Choice *D* is incorrect.

28. D: According to the phylogenetic tree in Figure 1, the common ancestor of the striped skunk and European otter is the one that is noted before they branch into separate lineages, which is Mustelidae. Mephitis, Choice *A*, is the genus for only the striped skunk. Felidae, Choice *B*, and Canidae, Choice *C*, are completely different branches of the Carnivora order than the one that leads to the striped skunk and European otter.

29. A: According to Figure 2, the common trait that is listed on the branch of the cladogram that leads to lizards and birds is seeing UV light. They also have the common traits listed on the main branch of the cladogram before their lineages are branched off, which are vertebrae, lungs, and amniotic eggs. Perch and flounder branch from the main common ancestor and develop spiny-rayed fins, Choice *B*. Lobsters and spiders branch from the main common ancestor and develop the ability to molt an exoskeleton, Choice *C*.

30. C: Domestic cats and wolves are proposed to be related at the point where they share a common line before any branching occurs to separate their lineages. Figure 1 shows this as Carnivora, which is noted as the Order on the left side of the figure.

31. D: The color reagent is attached to the secondary antibody. It is released only when the secondary antibody attaches to the activated primary antibody. The antigen, primary antibody, and capture antibody, Choices *A*, *B*, and *C*, do not have any color reagent attached to them, so only the secondary antibody can cause the color reaction.

32. B: The color reagent is attached to the secondary antibody. If more antigen is present, more primary and secondary antibody will be attached to it and more color reagent will be released. Row 1 has the darkest green color of all the samples tested in the plate in Figure 2.

33. C: Looking at the graph in Figure 3, the highest amount of IL-1β is found in the spleen. IL-1β is a marker of inflammation and indicates that the spleen had the most inflammation of the areas tested. Serum, Choice *A*, had no IL-1β in the sample. Plasma, Choice *B*, had the second highest amount of IL-1β in the sample, and bone marrow, Choice *D*, had the second lowest amount of IL-1β.

34. B: Looking at the diagram in Figure 1, the antigen is located between the capture antibody and the primary antibody. The capture antibody keeps the antigen attached to the surface of the plate. The primary antibody recognizes the specific antigen. The secondary antibody generally recognizes the primary antibody is not specific to the antigen.

35. A: ELISAs are used to analyze specific substances within a larger sample. The antibodies used in an ELISA are designed specifically for a particular antigen. Sandwich ELISAs are generally used to quantify one antigen and not all substances in a larger sample, making Choice *B* incorrect. When used to quantify an antigen, the antibodies need to already be developed and able to detect the antigen, making Choice *D* incorrect.

36. A: Stabilizing selection occurs when an intermediate phenotype is favored over two extreme phenotypes. In Scenario 1, the mice develop an intermediate colored fur so that they can blend in with the rocks in their environment. Developing one or both extremes, Choices *B* and *C*, would make them more visible to predators. Color selection, Choice *D*, is not a type of natural selection.

37. D: Directional selection occurs when one extreme of a phenotype is favored. In Scenario 2, large beaks are favored over medium- or small-sized beaks. The large beaks help the finches break up the tough seeds that became abundant after the drought. Finches with medium and small beaks had trouble breaking up the large seeds and did not survive as well as those with large beaks.

38. B: Small-beaked finches had trouble breaking up the large seeds after the drought, and therefore could not gain enough nutrition for survival. Natural selection is based on the idea that the individuals who adapt to their environment in the best way are the ones that have enhanced survival and reproduction. Finches with large beaks were most able to adapt to the large seeds and continue with their regular feeding schedule. There was no evidence of Choices *A* or *C* in the passage. Two extremes can be selected by natural selection in disruptive selection but that was not the case here, Choice *D*.

39. C: Disruptive selection occurs when both extremes of a phenotype are selected. In Cameroon, the finches had both large and small beaks, but did not survive well with medium beaks. If medium beaks were selected, it would have been stabilizing selection, Choice *A*. If only one of the extremes had been favored, it would have been directional selection, Choice *B*.

40. D: Natural selection is the idea that individuals are selected to survive and reproduce based on their ability to adapt to the environment. Their phenotypes are advantageous for survival and reproduction over those of other individuals. It is solely based on the phenotype of the individual, not the genotype, Choice *A*. Extreme phenotypes, Choice *B*, may be selected but are not always the most advantageous. It occurs all time, not just in extreme weather conditions, such as a drought, Choice *C*.

ACT Science Practice Test #3

Passage 1

Questions 1-5 pertain to the following information:

Worldwide, fungal infections of the lung account for significant mortality in individuals with compromised immune function. Three of the most common infecting agents are *Aspergillus, Histoplasma*, and *Candida*. Successful treatment of infections caused by these agents depends on an early and accurate diagnosis. Three tests used to identify specific markers for these mold species include ELISA (enzyme-linked immunosorbent assay), GM Assay (Galactomannan Assay), and PCR (polymerase chain reaction).

Two important characteristics of these tests include sensitivity and specificity. Sensitivity relates to the probability that the test will identify the presence of the infecting agent, resulting in a true positive result. Higher sensitivity equals fewer false-positive results. Specificity relates to the probability that if the test doesn't detect the infecting agent, the test is truly negative for that agent. Higher specificity equals fewer false-negatives.

Figure 1 shows the timeline for the process of infection from exposure to the pathogen to recovery or death.

Figure 1:
Natural History of the Process of Infection

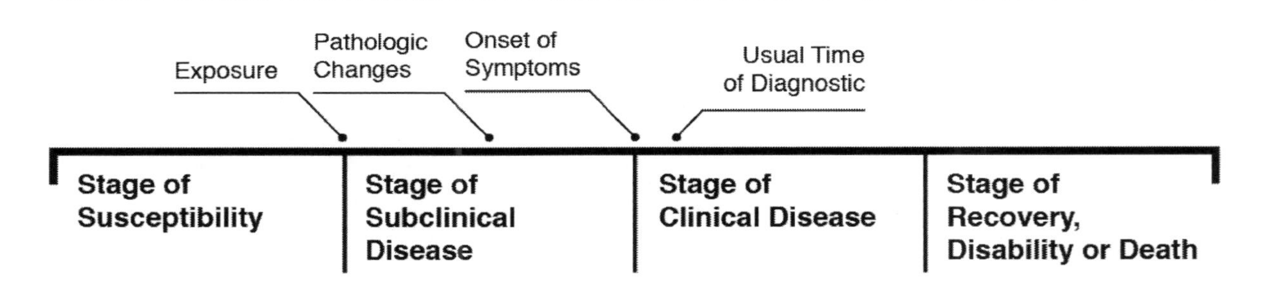

Figure 2 (below) shows the sensitivity and specificity for ELISA, GM assay and PCR related to the diagnosis of infection by *Aspergillus*, *Histoplasma* and *Candida*.

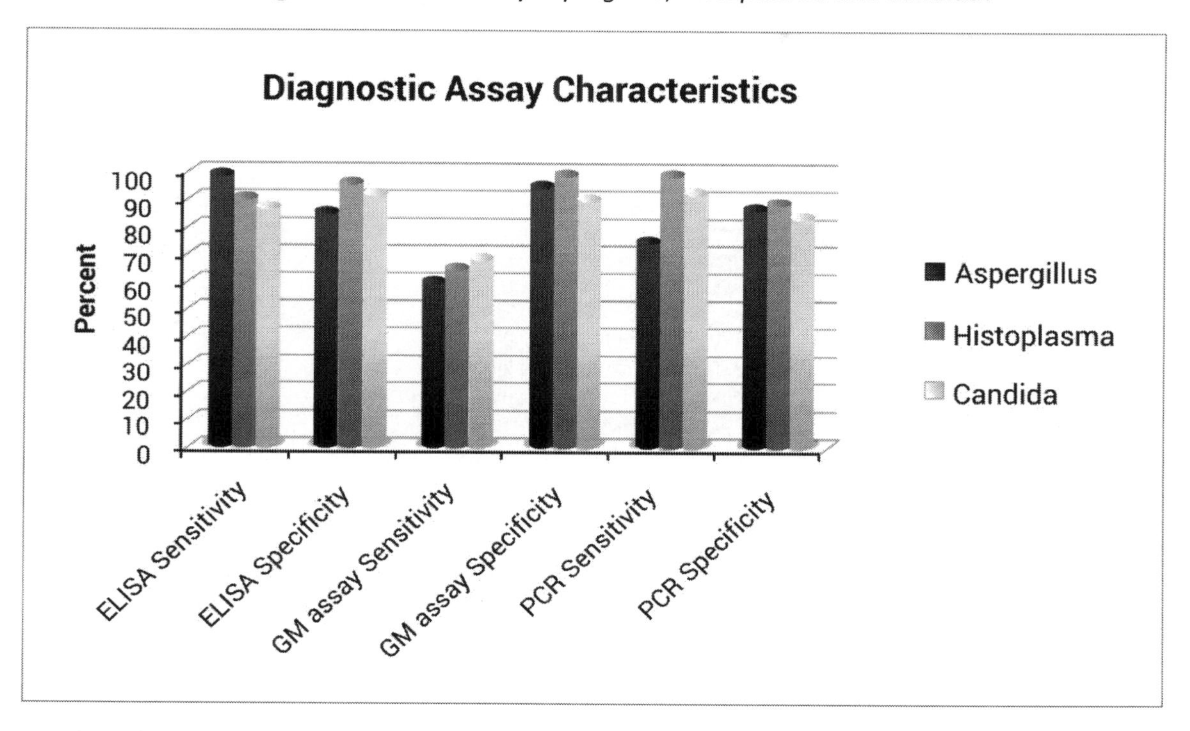

The table below identifies the process of infection in days from exposure for each of the species.

Process of Infection - Days Since Pathogen Exposure			
	Aspergillus	Histoplasma	Candida
Sub-clinical Disease	Day 90	Day 28	Day 7
Detection Possible	Day 118	Day 90	Day 45
Symptoms Appear	Day 145	Day100	Day 120

Figure 3 (below) identifies the point at which each test can detect the organism. Time is measured in days from the time an individual is exposed to the pathogen.

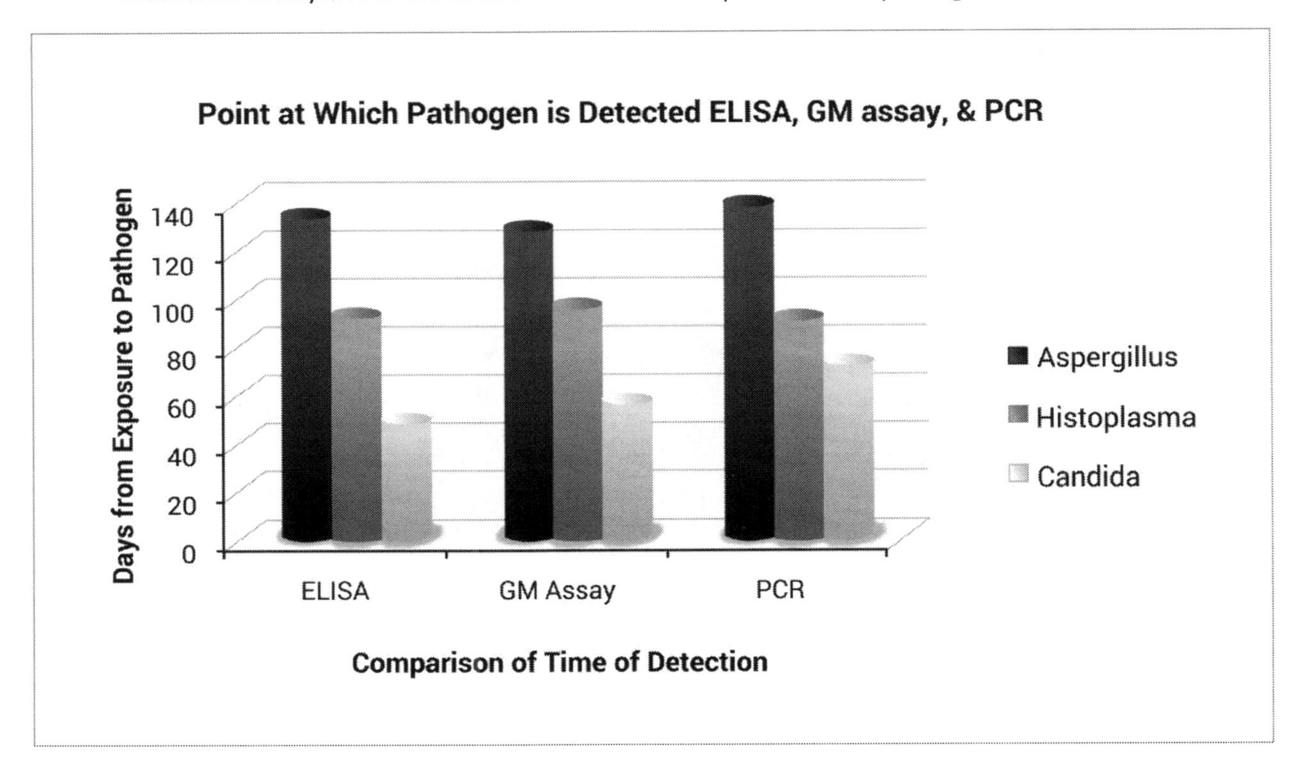

1. Which of the following statements is supported by Figure 2?
   a. For *Candida*, the GM assay will provide the most reliable results.
   b. ELISA testing for *Aspergillus* is the most specific of the three tests.
   c. PCR is the most sensitive method for testing *Histoplasma*.
   d. True positive rates were greater than 75% for all three testing methods.

2. In reference to the table and Figure 3, which pathogen can be detected earlier in the disease process, and by which method?
   a. *Candida* by PCR testing
   b. *Aspergillus* by ELISA testing
   c. *Candida* by GM assay
   d. *Histoplasma* by PCR testing

3. In reference to Figure 2, which statement is correct?
   a. There is a 20% probability that ELISA testing will NOT correctly identify the presence of *Histoplasma.*
   b. When GM assay testing for *Candida* is conducted, there is a 31% probability that it will NOT be identified if the organism is present.
   c. The probability that GM assay testing for *Aspergillus* will correctly identify the presence of the organism is 99%.
   d. The false-negative probabilities for each of the three testing methods identified in Figure 2 indicate that the organism will be detected when present less than 70% of the time.

4. Physicians caring for individuals with suspected *Histoplasma* infections order diagnostic testing prior to instituting treatment. PCR testing results will not be available for 10 days. GM assay results can be obtained more quickly. The physicians opt to wait for the PCR testing. Choose the best possible rationale for that decision.
   a. The treatment will be the same regardless of the test results.
   b. The individual was not exhibiting any disease symptoms.
   c. The probability of PCR testing identifying the presence of the organism is greater than the GM assay.
   d. The subclinical disease phase for *Histoplasma* is more than 100 days.

5. Referencing the data in Figures 2 and 3, if ELISA testing costs twice as much as PCR testing, why might it still be the best choice to test for *Candida*?
   a. ELISA testing detects the presence of *Candida* sooner than PCR testing.
   b. ELISA testing has fewer false-positives than PCR testing.
   c. There is only a 69% probability that PCR testing will correctly identify the presence of *Candida.*
   d. PCR testing is less sensitive than ELISA testing for *Candida.*

Passage 2

Questions 6-12 pertain to the following information:

Scientists disagree about the cause of Bovine Spongiform Encephalopathy (BSE), also known as "mad cow disease." Two scientists discuss different explanations about the cause of the disease.

Scientist 1

Mad cow disease is a condition that results in the deterioration of brain and spinal cord tissue. This deterioration manifests as sponge-like defects or holes that result in irreversible damage to the brain. The cause of this damage is widely accepted to be the result of an infectious type of protein, called a prion. Normal prions are located in the cell wall of the central nervous system and function to preserve the myelin sheath around the nerves. Prions are capable of turning normal proteins into other prions by a process that is still unclear, thereby causing the proteins to be "refolded" in abnormal and harmful configurations. Unlike viruses and bacteria, the harmful prions possibly don't contain DNA or RNA, based on the observation of infected tissues in the laboratory that remain infected after immersion in formaldehyde or exposure to ultraviolet light. The transformation from normal to abnormal protein structure and function in a given individual is thought to occur as the result of proteins that are genetically weak or abnormally prone to mutation, or through transmission from another host through food, drugs or organ transplants from infected animals. The abnormal prions also don't trigger an immune response. After prions accumulate in large enough numbers, they form damaging conglomerations that result in the sponge-like holes in tissues, which eventually cause the loss of proper brain function and death.

Figure 1 depicts formation of abnormal prions that results from the abnormal (right) folding of amino acids.

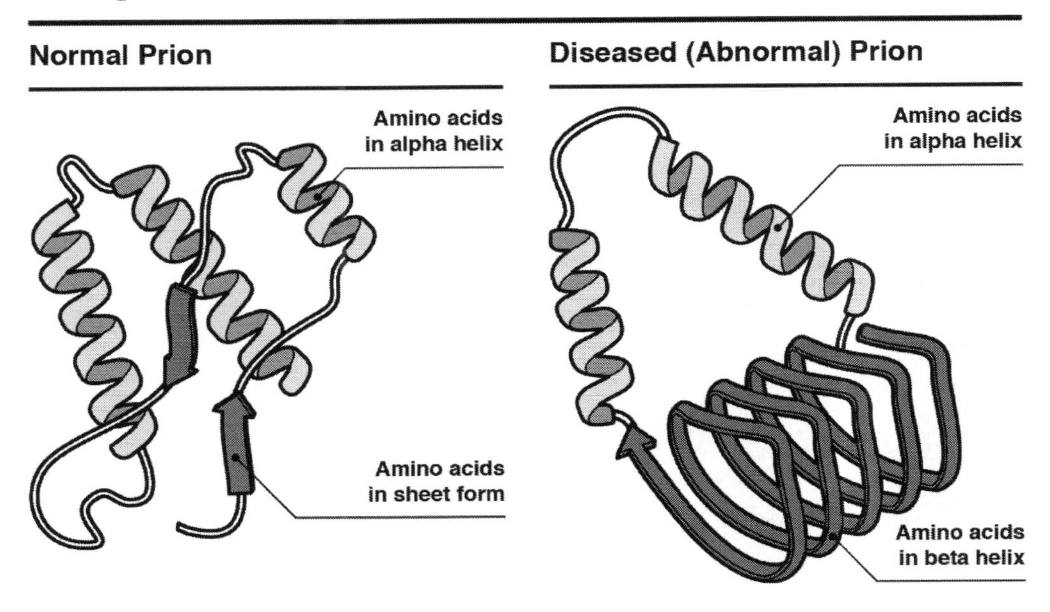

Scientist 2

The degeneration of brain tissue in animals afflicted with mad cow disease is widely considered to be the result of prion proteins. This theory fails to consider other possible causes, such as viruses. Recent studies have shown that infected tissues often contain small particles that match the size and density of viruses. In order to demonstrate that these viral particles are the cause of mad cow disease, researchers used chemicals to inactivate the viruses. When the damaged, inactivated viruses were introduced into healthy tissue, no mad cow disease symptoms were observed. This result indicates that viruses are likely the cause of mad cow disease. In addition, when the infected particles from an infected animal are used to infect a different species, the resulting particles are identical to the original particles. If the infecting agent was a protein, the particles would not be identical because proteins are species-specific. Instead, the infective agent is viewed as some form of a virus that has its own DNA or RNA configuration and can reproduce identical infective particles.

6. Which statement below best characterizes the main difference in the scientists' opinions?
   a. The existence of species-specific proteins
   b. Transmission rates of mad cow disease
   c. The conversion process of normal proteins into prions
   d. The underlying cause of mad cow disease

7. Which of the following statements is INCORRECT?
   a. Scientist 2 proposes that viruses aren't the cause of mad cow disease because chemicals inactivated the viruses.
   b. Scientist 1 suggests that infectious proteins called prions are the cause of mad cow disease.
   c. Scientist 1 indicates that the damaging conglomerations formed by prions eventually result in death.
   d. Scientist 2 reports that infected tissues often contain particles that match the size profile of viruses.

8. which of the following is true according to Scientist 1?
   a. Normal proteins accumulate in large numbers to produce damaging conglomerations.
   b. Prions can change normal proteins into prions.
   c. Species-specific DNA sequences of infected tissues indicate that proteins cause mad cow disease.
   d. Prions are present only in the peripheral nervous system of mammals.

9. Which of the following statements would be consistent with the views of BOTH scientists?
   a. Resulting tissue damage is reversible.
   b. The infecting agent is composed of sheets of amino acids in an alpha helix configuration.
   c. Species-specific DNA can be isolated from infected tissue.
   d. Cross-species transmission of the illness is possible.

10. How does the "conglomeration" described in the passage affect function?
    a. Synapses are delayed
    b. Sponge-like tissue formations occur
    c. Space-occupying lesions compress the nerves
    d. The blood supply to surrounding tissues is decreased

11. What evidence best supports the views of Scientist 2?
    a. Species-specific DNA is present in the infected particles.
    b. Prions are present in the cell membrane.
    c. Prions can trigger an immune response.
    d. The infected particles were inactivated and didn't cause disease.

12. Which of the following statements is supported by this passage?
    a. Scientist 1 favors the claim that viruses are the cause of mad cow disease.
    b. Prions are a type of infectious virus.
    c. The process that results in the formation of the abnormal prion is unclear.
    d. Mad cow disease is caused by normal proteins.

Passage 3

Questions 13-17 pertain to the following information:

> Scientists have long been interested in the effect of sleep deprivation on overeating and obesity in humans. Recently, scientists discovered that increased levels of the endocannabinoid 2-Arachidonoylglycerol (2-AG) in the human body is related to overeating. The endocannabinoids play an important role in memory, mood, the reward system, and metabolic processes including glucose metabolism and generation of energy. The endocannabinoid receptors CB1-R and CB2-R are protein receptors located on the cell membrane in the brain, the spinal cord and, to a lesser extent, in the

peripheral neurons and the organs of the immune system. The two principal endogenous endocannabinoids are AEA (Anandamide) and 2-Arachidonoylglycerol (2-AG). The endocannabinoids can affect the body's response to chronic stress, mediate the pain response, decrease GI motility, and lessen the inflammatory response in some cancers.

Figure 1 (below) identifies the chemical structure of the endogenous cannabinoids including 2-AG.

Figure 1:

Chemical Structure of Common Endogenous Cannabinoids

The Five-Best known Endocannabinoids Showing the Common 19 - C Backbone Structure and specific R-group Constituents

Anandamide

2 - Arachidonoyl- glycerol

Noladin Ether

N-arachidonoyl-dopamine

Virodhamine

EC backbone structure

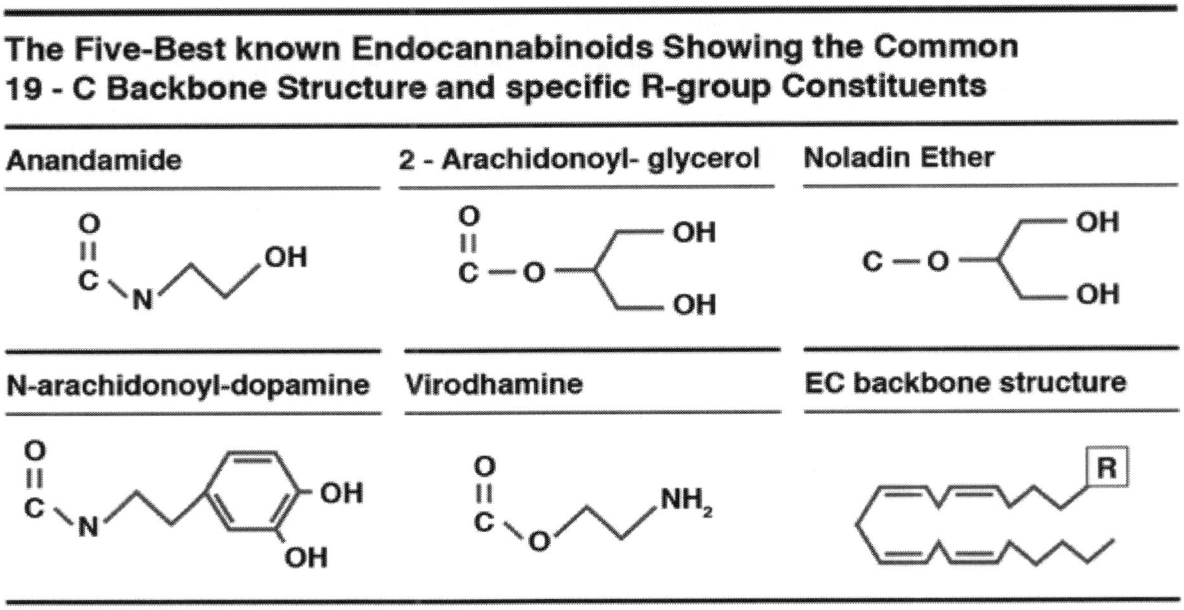

Recent research has also examined the relationship between sleep deprivation and the levels of 2-AG present in blood, as these conditions relate to obesity. The circadian fluctuations of 2-AG are well-known. Levels normally increase in late afternoon and evening. This physiological increase is thought to contribute to late-day snacking behaviors even after adequate calories have been consumed. The relationship between sleep deprivation and 2-AG appears to relate to the effect of 2-AG on the stress response, represented by sleep deprivation in this study. In order to examine this relationship, university scientists conducted an experiment to identify the influence of injections of 2-AG and sleep deprivation on overeating in a population of non-obese male and female participants that ranged in age from 20 – 40 years old. To accomplish this, human research subjects (participants) were allowed to eat their favorite junk foods in addition to consuming sufficient calories each day. All of the participants were injected daily with a solution of either sterile normal saline or 2-AG. Daily weight gain was recorded for the three treatment groups that included: participants A - E who received sterile normal saline injections, participants F - J who received 2-AG injections, and participants K - O who received 2-AG injections and were limited to 4.5 hours of sleep each night for 7 nights. The results of the three trials are shown below.

Figure 2 identifies the daily weight gain (in grams) of participants receiving sterile normal saline injections.

Daily Weight Gain for Patients Receiving Sterile Normal Saline Injections

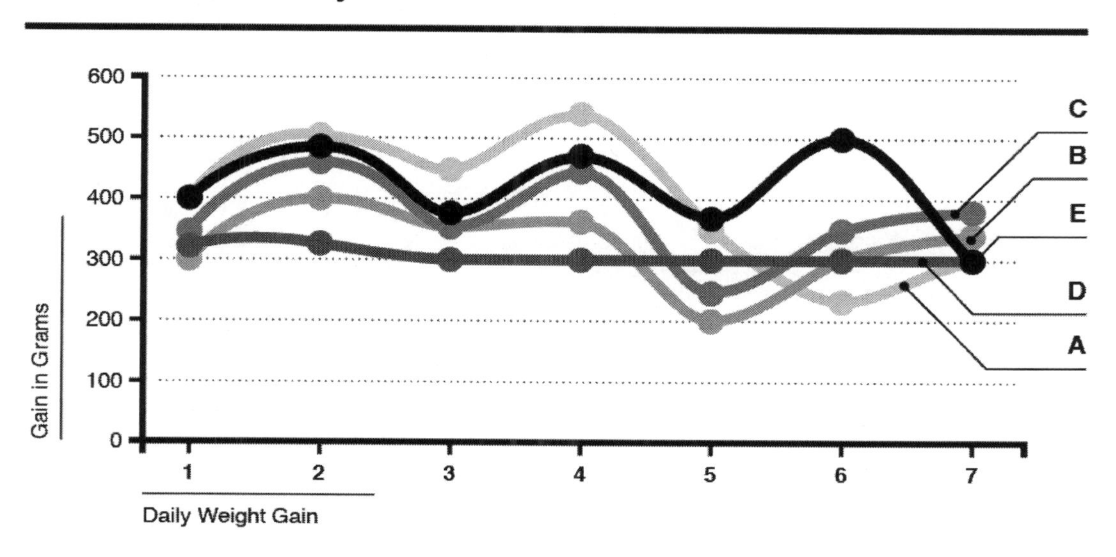

Figure 3 identifies the daily weight gain for participants receiving 2-AG injections.

Figure 3:
Daily Weight Gain for Participants Receiving Daily 2-AG Injections

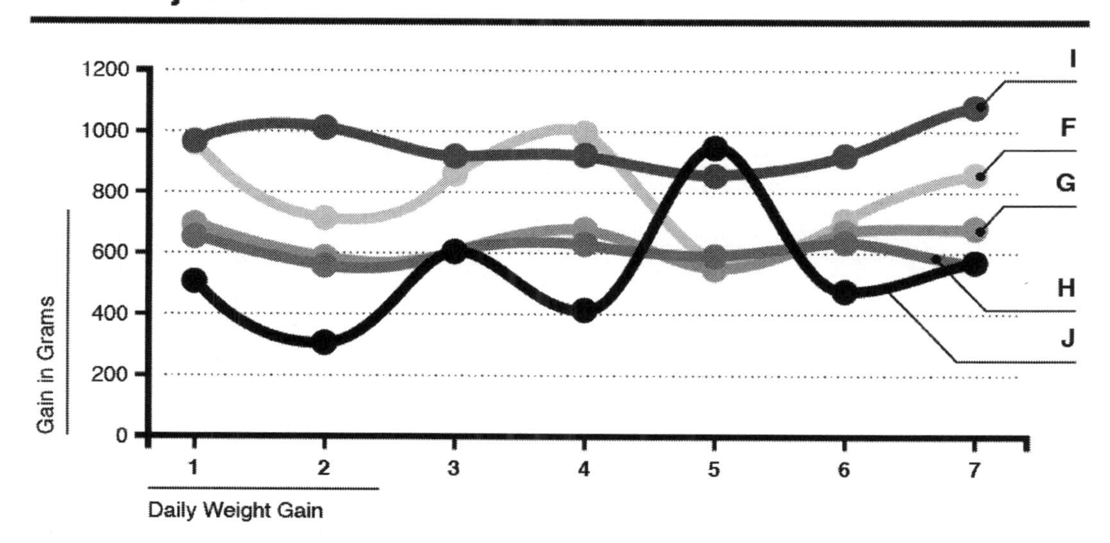

Figure 4 identifies the daily weight gain for participants receiving daily injections of 2-AG who were also limited to 4.5 hours sleep per night for 7 consecutive nights.

Figure 4:

Daily Weight Gain for Participants Receiving Daily 2-AG Injections Who Were Limited to 4.5 Hours of Sleep Per Night for 7 Consecutive Nights

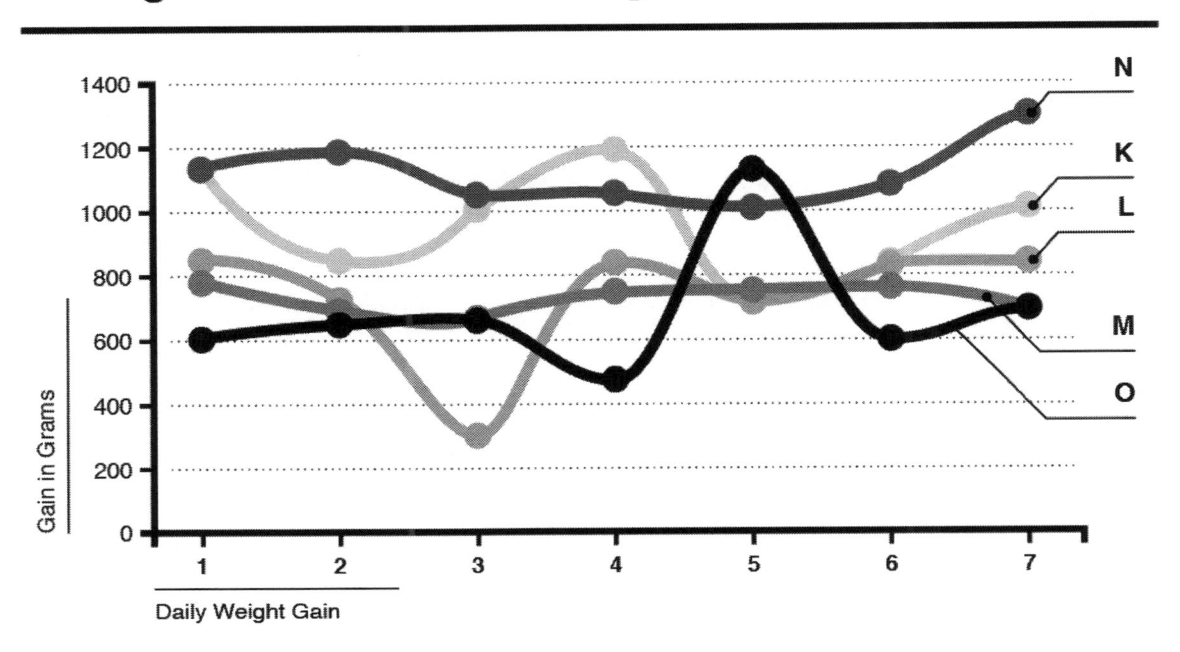

Figure 5 (below) identifies the participants' average daily weight gain by trial.

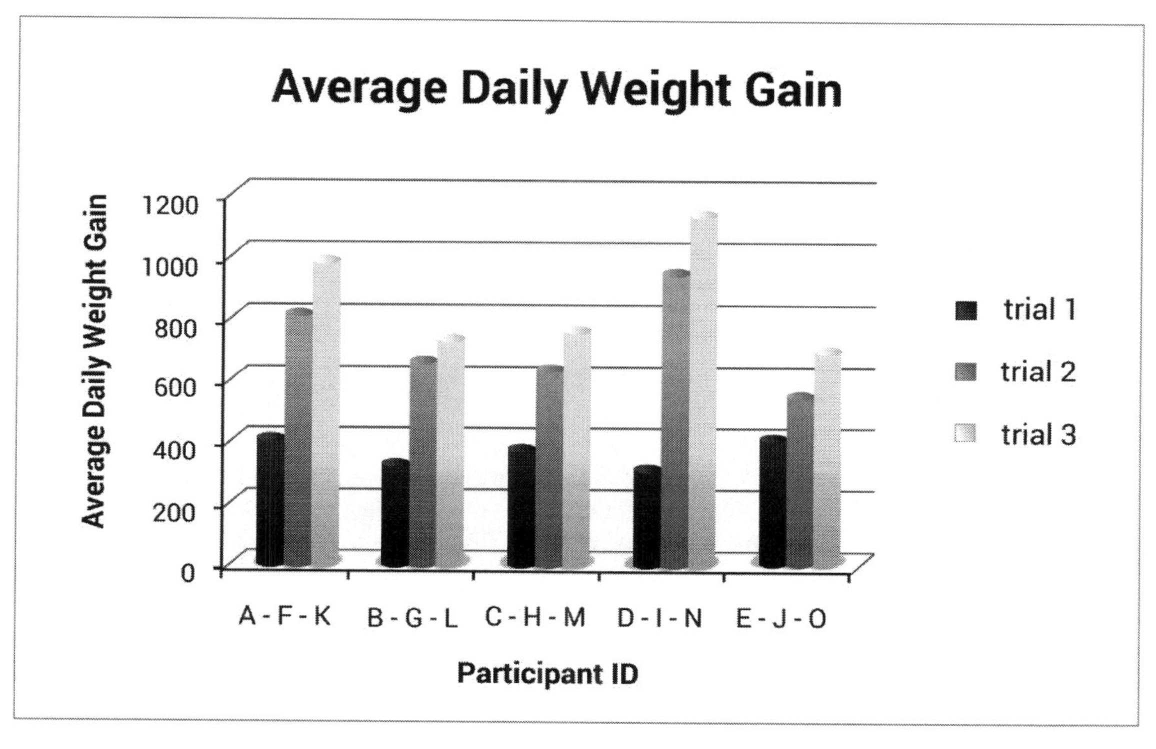

13. What was the main hypothesis for this study?
   a. 2-AG injections combined with sleep deprivation will result in weight gain.
   b. 2-AG injections will increase food intake beyond satiety.
   c. Sleep deprivation will result in weight gain.
   d. The placebo effect of the sterile normal saline will influence eating behavior.

14. Do the study results support the hypothesis? Choose the best answer.
   a. No, participants in trials 1 and 3 all gained weight.
   b. Yes, participants in trial 1 gained more weight daily than participants in trial 3.
   c. No, the average weight gain of participants in trial 2 and trial 3 was the same.
   d. Yes, all trial 3 participants gained more weight than trial 1 participants.

15. Describe the study results for participants D and H.
   a. Participant H gained more than one pound each day.
   b. Weight gain for each participant was inconsistent with the study hypothesis.
   c. There was significant fluctuation in the daily weight gain for both participants.
   d. Participant D's average daily weight was two times participant H's average daily weight gain.

16. According to the researchers, which of the following best describes the influence of sleep deprivation on eating behaviors?
   a. The total number of sleep hours is unrelated to the degree of body stress.
   b. Sleep deprivation stimulates the release of endogenous cannabinoids that may increase food intake.
   c. Deprivation of any variety triggers the hunger response.
   d. Sleep deprivation increases eating behaviors in the early morning hours.

17. According to the passage, how does 2-AG influence eating behaviors?
    a. Circadian fluctuations result in increased levels of 2-AG in the afternoon and evening.
    b. Endogenous cannabinoids like 2-AG increase gastric motility, which stimulates the hunger response.
    c. The sedation that results from the presence of 2-AG limits food intake.
    d. Endogenous cannabinoids block the opioid system, which decreases food-seeking behaviors.

Passage 4

Questions 18-22 pertain to the following passage:

A national wholesale nursery commissioned research to conduct a cost/benefit analysis of replacing existing fluorescent grow lighting systems with newer LED lighting systems. LEDs (light-emitting diodes) are composed of various semi-conductor materials that allow the flow of current in one direction. This means that LEDs emit light in a predictable range, unlike conventional lighting systems that give off heat and light in all directions. The wavelength of light of a single LED is determined by the properties of the specific semi-conductor. For instance, the indium gallium nitride system is used for blue, green, and cyan LEDs. As a result, growing systems can be individualized for the specific wavelength requirements for different plant species. In addition, LEDs don't emit significant amounts of heat compared to broadband systems, so plant hydration can be controlled more efficiently.

Figure 1 identifies the visible spectrum with the wavelength expressed in nanometers.

Figure 1:
The Visible Spectrum (Wavelength in Nanometers)

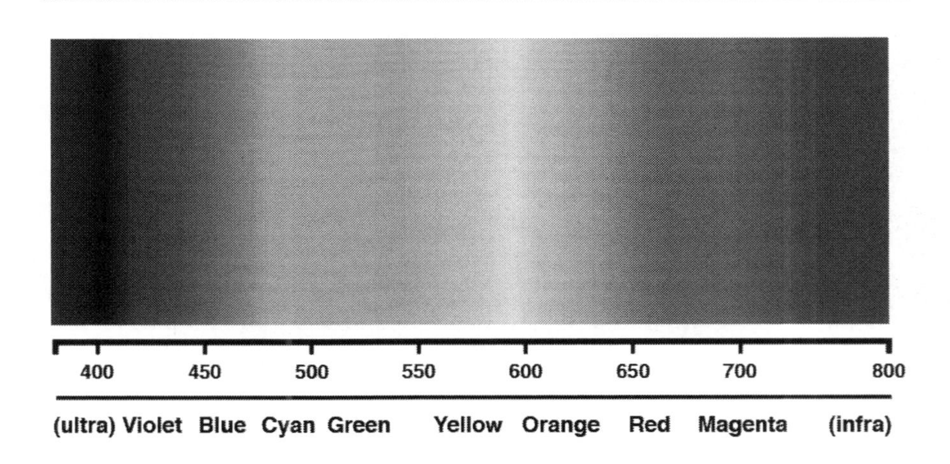

Figure 2 (below) identifies the absorption rates of different wavelengths of light.

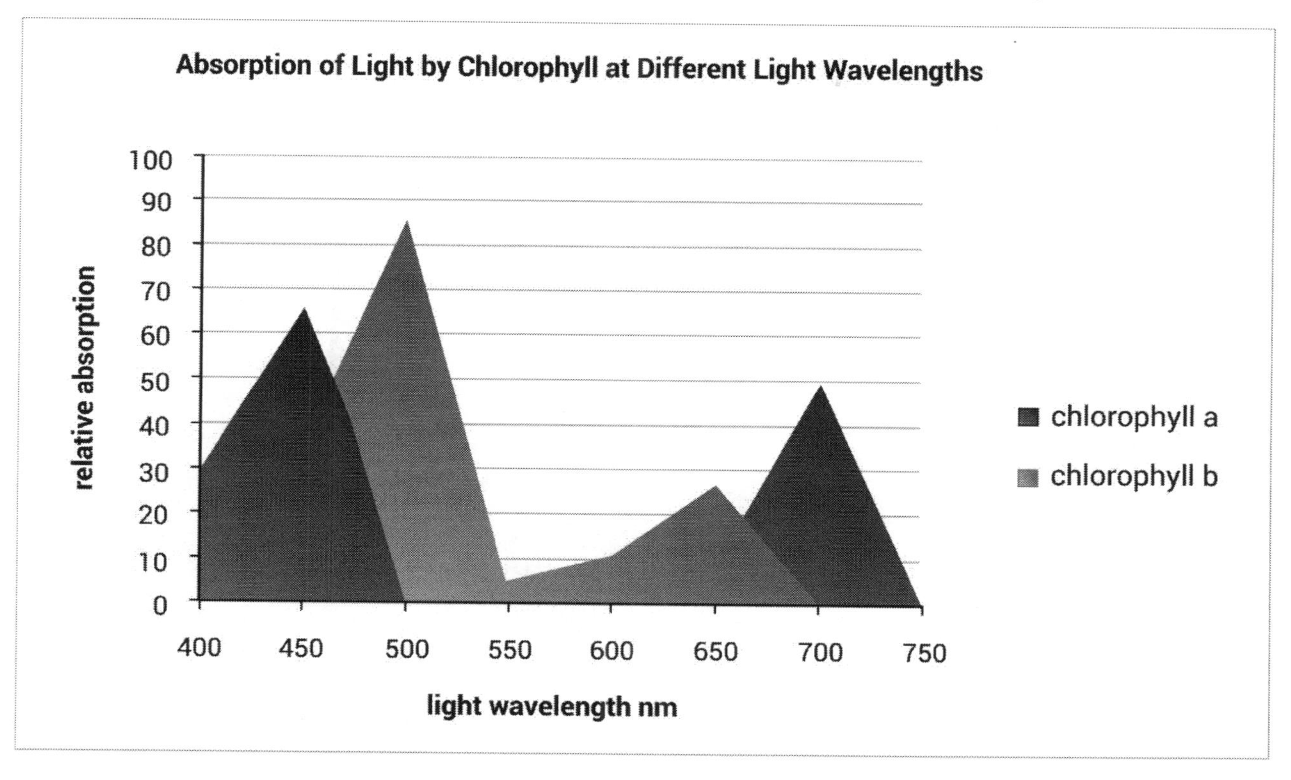

Researchers conducted three trials and hypothesized that LEDs would result in greater growth rates than conventional lighting or white light. They also hypothesized that using a combination of red, blue, green, and yellow wavelengths in the LED lighting system would result in a greater growth rate than using red or blue wavelengths alone. Although green and yellow wavelengths are largely reflected by the plant (Figure 2), the absorption rate is sufficient to make a modest contribution to plant growth. Fifteen Impatiens walleriana seed samples were planted in the same growing medium. Temperature, hydration, and light intensity were held constant. Plant height in millimeters was recorded as follows.

Figure 3 identifies the plant growth rate in millimeters with light wavelengths of 440 nanometers.

Figure 3:
Plant Growth Rate (mm) with Light wavelengths of 440 nm

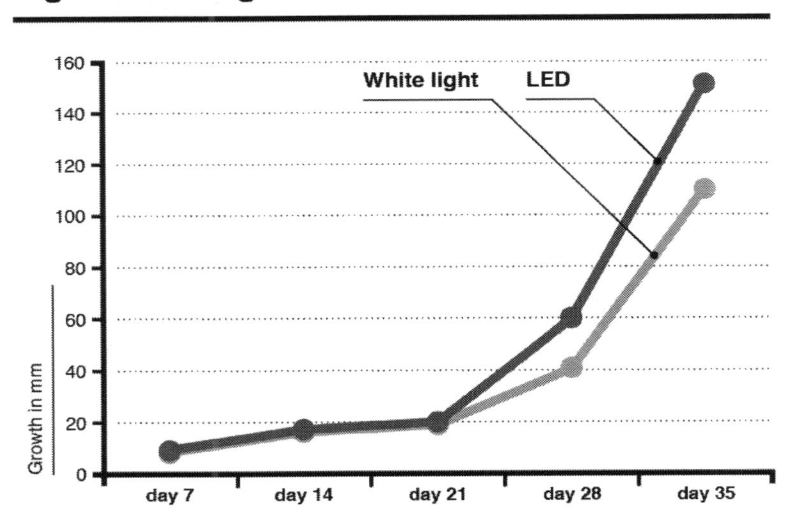

Figure 4 (below) identifies the plant growth rate in mm with light wavelengths of 650 nanometers.

Figure 4:
Plant Growth Rate (mm) with Light wavelengths of 650 nm

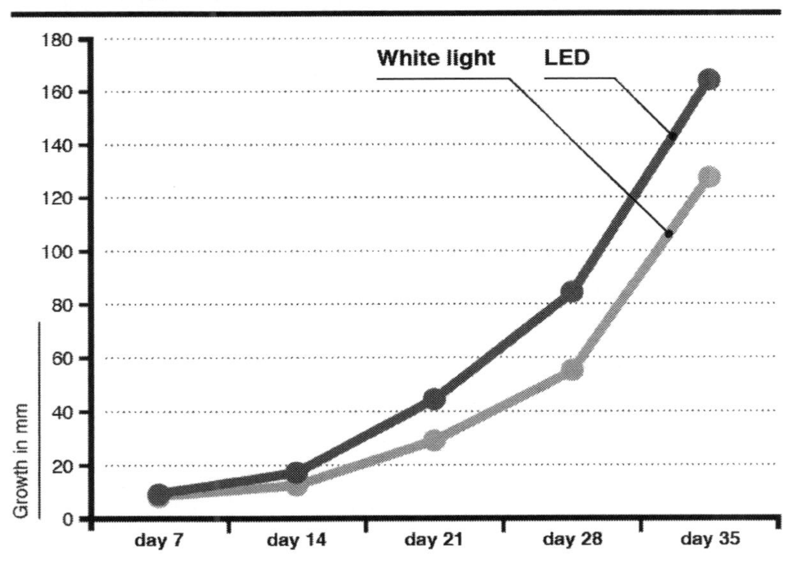

Figure 5 (below) identifies the plant growth rate in millimeters with combined light wavelengths of 440, 550, and 650 nanometers.

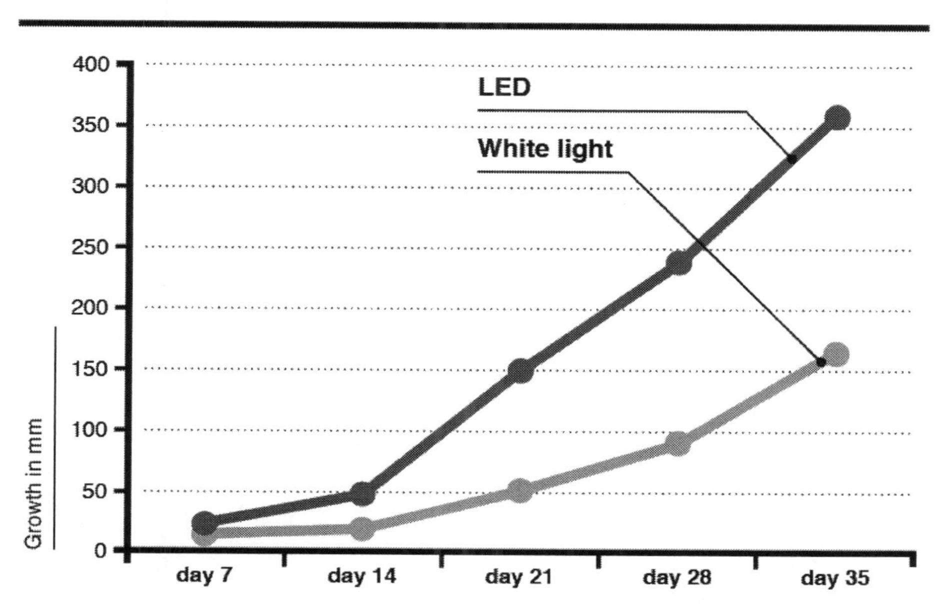

Figure 6 (below) identifies average daily plant growth rate in millimeters.

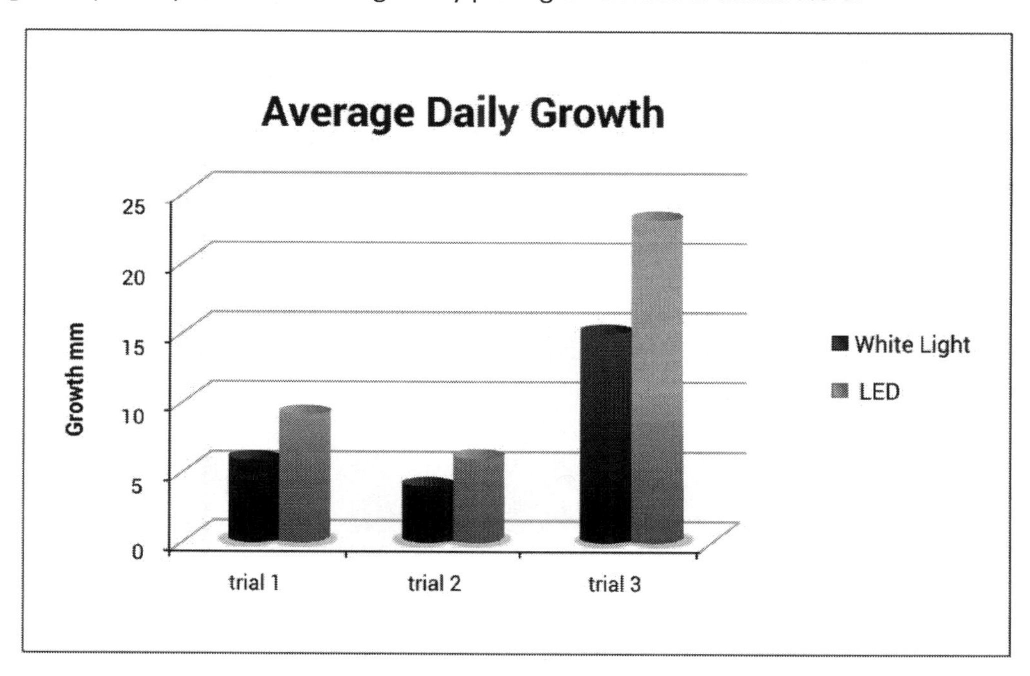

18. If the minimum plant height required for packaging a plant for sale is 150 millimeters, based on plant growth, how much sooner will the LED plants be packaged compared to the white light plants?
   a. 14 days
   b. 21 days
   c. 35 days
   d. 42 days

19. Plants reflect green and yellow light wavelengths. Do the results of the three trials support the view that plants also absorb and use green and yellow light wavelengths for growth?
   a. Yes, green and yellow light wavelengths were responsible for plant growth in trial 3.
   b. No, white light alone was responsible for measurable plant growth.
   c. Yes, the growth rates in trial 3 were greater than the rates in trials 1 and 2.
   d. No, only the red and blue wavelengths were effective in stimulating plant growth.

20. When did the greatest rate of growth occur for both groups in trial 1 and trial 2?
   a. From 7 days to 14 days
   b. From 28 days to 35 days
   c. From 21 days to 28 days
   d. From 14 days to 21 days

21. If an LED lighting system costs twice as much as a white light system, based only on the average daily growth rate as noted above, would it be a wise investment?
   a. No, because multiple different semi-conductors would be necessary.
   b. Yes, growth rates are better with LEDs.
   c. No, the LED average daily growth rate was not two times greater than the white light rate.
   d. Yes, LEDs use less electricity and water.

22. If the researchers conducted an additional trial, trial 4, to measure the effect of green and yellow wavelengths on plant growth, what would be the probable result?
   a. The growth rate would exceed trial 1.
   b. The growth rate would equal trial 3.
   c. The growth rate would be the same as trial 2.
   d. The growth rate would be less than trial 1 or trial 2.

Passage 5

Questions 23-28 pertain to the following passage:

> Mangoes are a tropical fruit that grow on trees native to Southern Asia called the *Mangifera*. Mangoes are now grown in most frost-free tropical and subtropical locations around the world. India and China harvest the greatest numbers of mangoes. A major problem the mango industry faces each year is the destruction of fruit after harvest. This destruction is the result of spoilage or rotting that occurs during long shipping and storage times.
>
> To prevent the spoilage of mangoes, fruits are stored and shipped in climate-controlled containers. Ideally, mangoes should be stored at around 5 °C, which is about the same temperature as a home refrigerator. Although storage at 5 °C is highly effective at

preventing spoilage, the monetary costs associated with maintaining this temperature during long shipping times are prohibitive.

Fruit companies spend large amounts of money to learn about the underlying cause of spoilage and possible methods to prevent loss of their product. Anthracnose, an infection that causes mango decay, is caused by *Colletotrichum,* a type of fungus that has been identified as a major contributor to mango spoilage. This fungus, which may remain dormant on green fruit, grows on the surface of the mango and can penetrate the skin and cause spoilage. The infection first appears during the flowering period as small black dots that progress to dark brown or black areas as ripening occurs. Humidity and excessive rainfall increase the severity of this infection. Previous studies established that colony sizes smaller than 35 millimeters after 4 weeks of travel resulted in acceptable amounts of spoilage.

Currently, several additional pre-treatment measures aimed at prevention are employed to slow decay of the fruit from the harvest to the marketplace. Industry researchers examined the individual and collective benefits of two of these processes, including post-harvest hot water treatment and air cooling at varied transport temperatures in order to identify optimum post-harvest procedures.

Table 1 identifies the observed mango decay in millimeters at 5 °C, 7.5 °C, and 10 °C with two pre-treatment processes over time measured in days since harvest of the fruit.

Table 1: Days Since Harvest

	2	**4**	**6**	**8**	**10**	**12**	**14**	**16**	**18**	**20**	**22**	**24**	**26**	**28**
5° C														
Water	1	4	7	9	11	12	14	19	22	23	25	27	28	30
Air	0	2	3	6	8	9	11	12	13	15	16	17	18	19
7.5°C														
Water	2	3	4	5	6	8	9	11	12	13	14	15	16	27
Air	0	2	5	6	7	9	10	15	22	23	24	27	32	39
10°C														
Water	2	3	5	7	8	11	12	14	22	35	42	44	47	62
Air	1	2	4	6	7	9	10	15	19	23	27	29	35	44

Figure 1 (below) identifies the observed mango decay of fruit stored at 5 °C measured in millimeters, with two pretreatment processes, over time measured in days since harvest of the fruit.

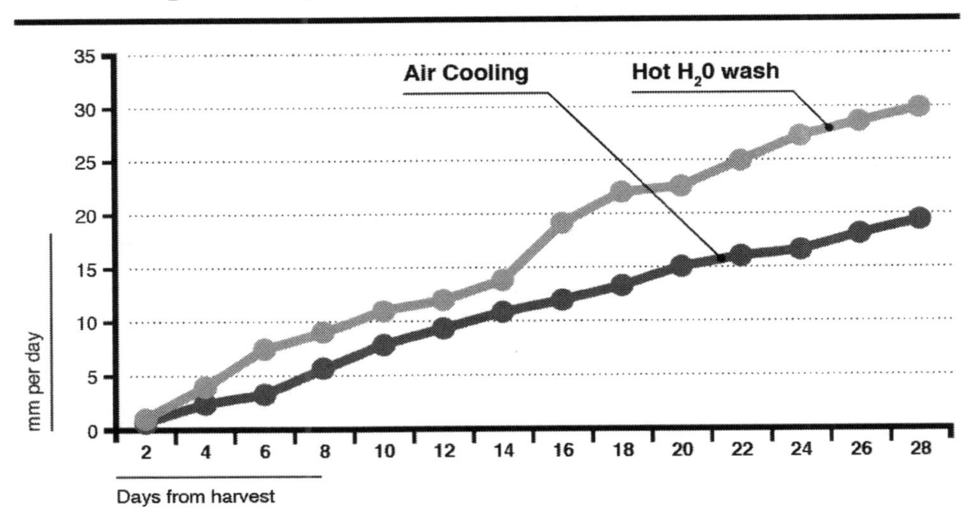

Figures 2 (below) identifies the observed mango decay of fruit stored at 7.5 °C, measured in millimeters, with two pretreatment processes, over time measured in days since harvest of the fruit.

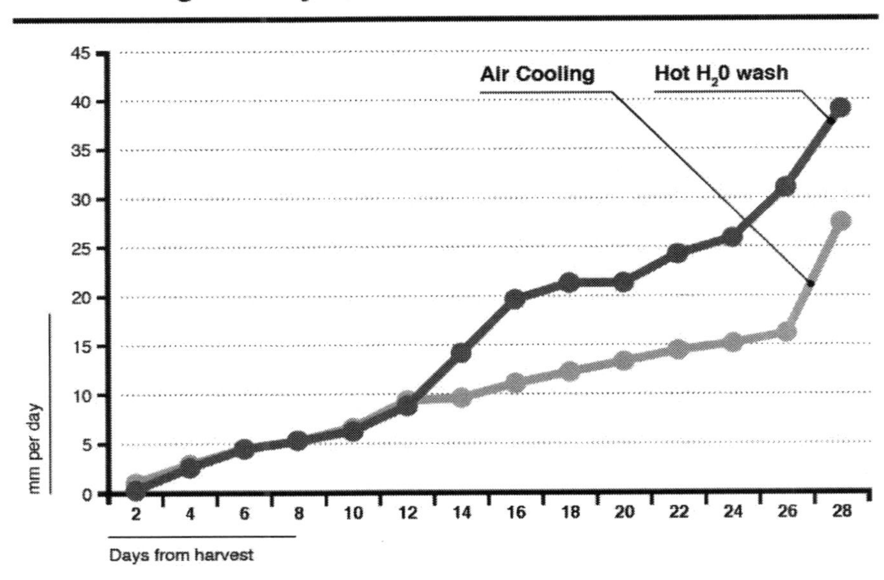

Figures 3 (below) identifies the observed mango decay of fruit stored at 10 °C measured in millimeters, with two pretreatment processes, over time measured in days since harvest of the fruit.

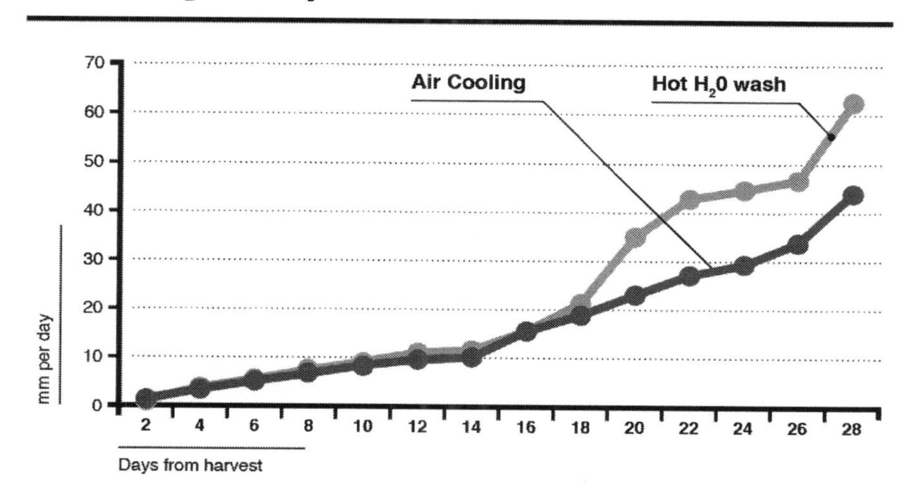

Figure 4 (below) identifies fruit decay measured in millimeters at 5 °C, 7.5 °C, and 10 °C with the combined pre-treatments over 28 days.

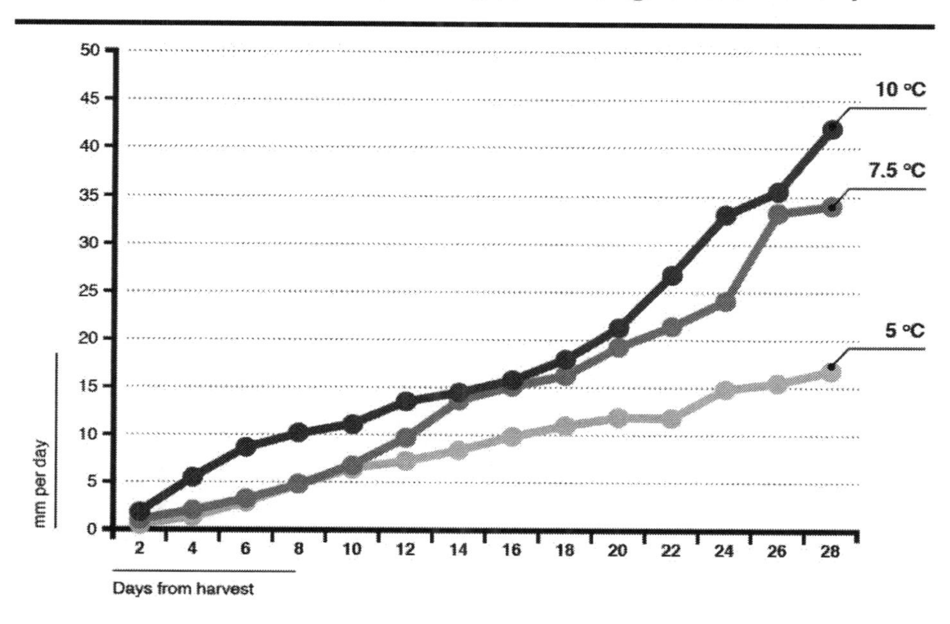

23. According to the passage above, which of the following statements is false?
    a. The optimal temperature for storing mangos is 5 °C.
    b. Anacardiaceae Magnifera is responsible for mango spoilage.
    c. Storing fruit at 5 °C is costly.
    d. Long distance shipping is a critical factor in mango spoilage.

24. If the mangoes were shipped from India to the U.S., and the trip was expected to take 20 days, which model would be best according to the data in Table 1?
    a. The 10 °C model, because fungal levels were acceptable for both pre-treatments.
    b. The 5 °C, model because it's more cost-effective.
    c. The 7.5 °C model, because this temperature is less expensive to maintain, and the fungal levels were acceptable.
    d. No single model is better than the other two models.

25. According to Figures 1 – 3 above, the largest one-day increase in fruit decay occurred under which conditions?
    a. Air cooling at 10 °C
    b. Hot water wash at 10 °C
    c. Air cooling at 7.5 °C
    d. Hot water wash at 7.5 °C

26. Which pre-treatment method reached unacceptable fungal levels first?
    a. Hot water wash at 7.5 °C
    b. Air cooling at 5 °C
    c. Hot water wash at 10 °C
    d. Air cooling at 10 °C

27. The researchers were attempting to identify the best shipping conditions for mangoes for a 28-day trip from harvest to market. Referencing Figures 1 - 4, which conditions would be the most cost-effective?
    a. Air cooling pre-treatment at 5 °C
    b. Air cooling and hot water wash pre-treatment at 5 °C
    c. Hot water wash pre-treatment at 7.5 °C
    d. Air cooling at 10 °C

28. Shipping mangoes at 5 °C is costly. According to the researchers' findings, is shipping mangoes at 5 °C more cost effective than 7.5 °C for trips lasting more than 28 days when combined air cooling and hot water wash treatments are applied?
    a. Yes, shipping at 7.5 °C combined with both pre-treatments resulted in an unacceptable fungal infection rate.
    b. Yes, fungal infection rates were below 35 mm for both pre-treatments 5 °C.
    c. No, air cooling pre-treatment was acceptable at 10 °C, and it's less expensive to ship fruit at 10 °C.
    d. No, hot water wash rates were lower than air cooling at 5 °C.

Passage 6

Questions 29-34 pertain to the following passage:

Scientists recently discovered that circadian rhythms help regulate sugar consumption by brown adipose tissue. The results of this study suggest that circadian rhythms and fat cells work together to warm the body in preparation for early morning activities involving cold weather. A circadian rhythm refers to life processes controlled by an internal "biological clock" that maintains a 24-hour rhythm. Sleep is controlled by one's circadian rhythm. To initiate sleep, the circadian rhythm stimulates the pineal gland to release the hormone melatonin, which causes sleepiness. Importantly, the circadian rhythm discerns when to begin the process of sleep based on the time of day. During the daytime, sunlight stimulates special cells within the eye, photosensitive retinal ganglion cells, which, in turn, allow the "biological clock" to keep track of how many hours of sunlight there are in a given day.

Brown adipose tissue (BAT) is a type of fat that plays an important role in thermogenesis, a process that generates heat. In humans and other mammals, there are two basic types of thermogenesis: shivering thermogenesis and non-shivering thermogenesis. Shivering thermogenesis involves physical movements, such as shaky hands or clattering teeth. Heat is produced as a result of energy being burned during physical activity. Non-shivering thermogenesis doesn't require physical activity; instead, it utilizes brown adipose tissue to generate heat. Brown fat cells appear dark because they contain large numbers of mitochondria, the organelles that burn sugar to produce energy and heat.

Researchers know that brown adipose tissue (BAT) is essential for maintaining body temperature. A new discovery in humans has shown that circadian rhythms cause BAT to consume more sugar in the early-morning hours. This spike in sugar consumption causes more heat to be produced in BAT. Scientists propose that our human ancestors could have benefited from extra body heat during cold hunts in the morning.

Perhaps more significantly, these new findings may suggest a role for BAT in the prevention of Type 2 Diabetes. Two important questions remain; to what degree does BAT affect blood glucose levels, and is it possible to increase BAT in a given individual? The demonstrated increase in sugar consumption and heat production of BAT is thought to be related to insulin-sensitivity. To examine the first question, researchers conducted three trials to examine the relationship between brown fat and blood glucose levels at different points in the day. PET scanning was used to estimate total body brown fat in 18 non-diabetic participants. Total body brown fat expressed as a proportion of total body fat (either 5%, 10%, or 20%) was the basis for group selection. The researchers hypothesized that the blood glucose levels would be inversely related to the percentage of BAT. Resulting data is included below.

Figure 1 (below) identifies the circadian cycle of blood glucose.

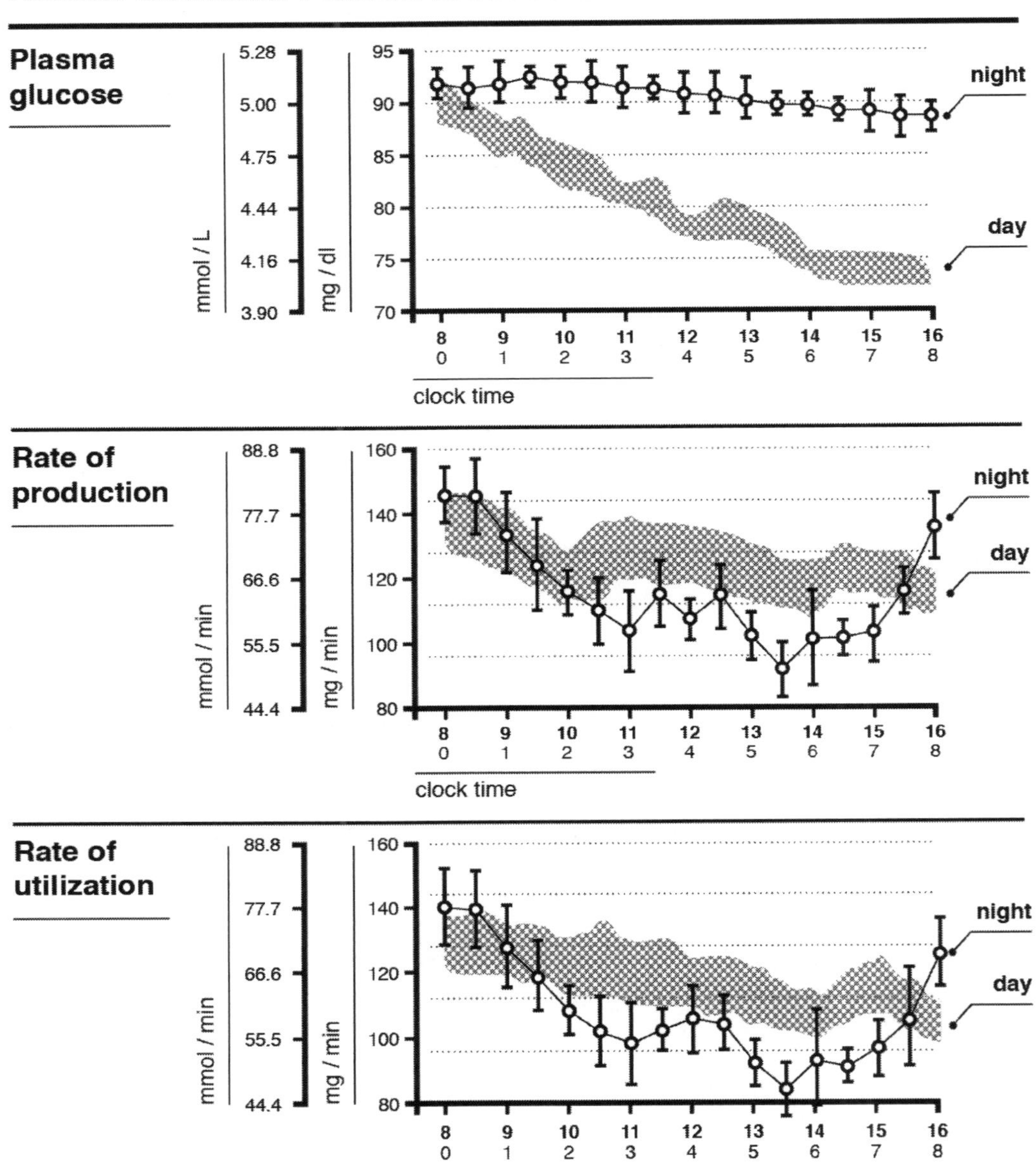

Figure 2 (below) identifies the resulting blood glucose measurements for participants with 5% total brown body fat.

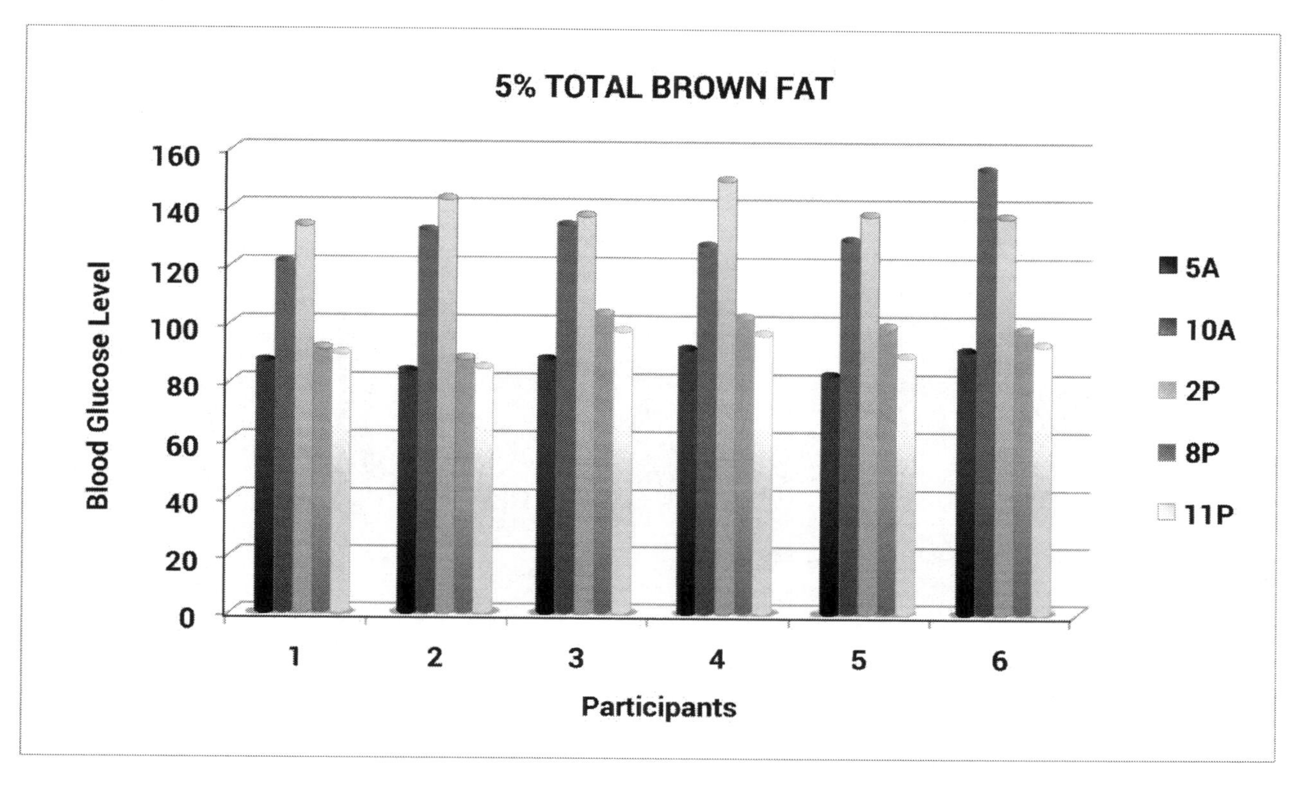

Figure 3 (below) identifies the resulting blood glucose measurements for participants with 10% total brown body fat.

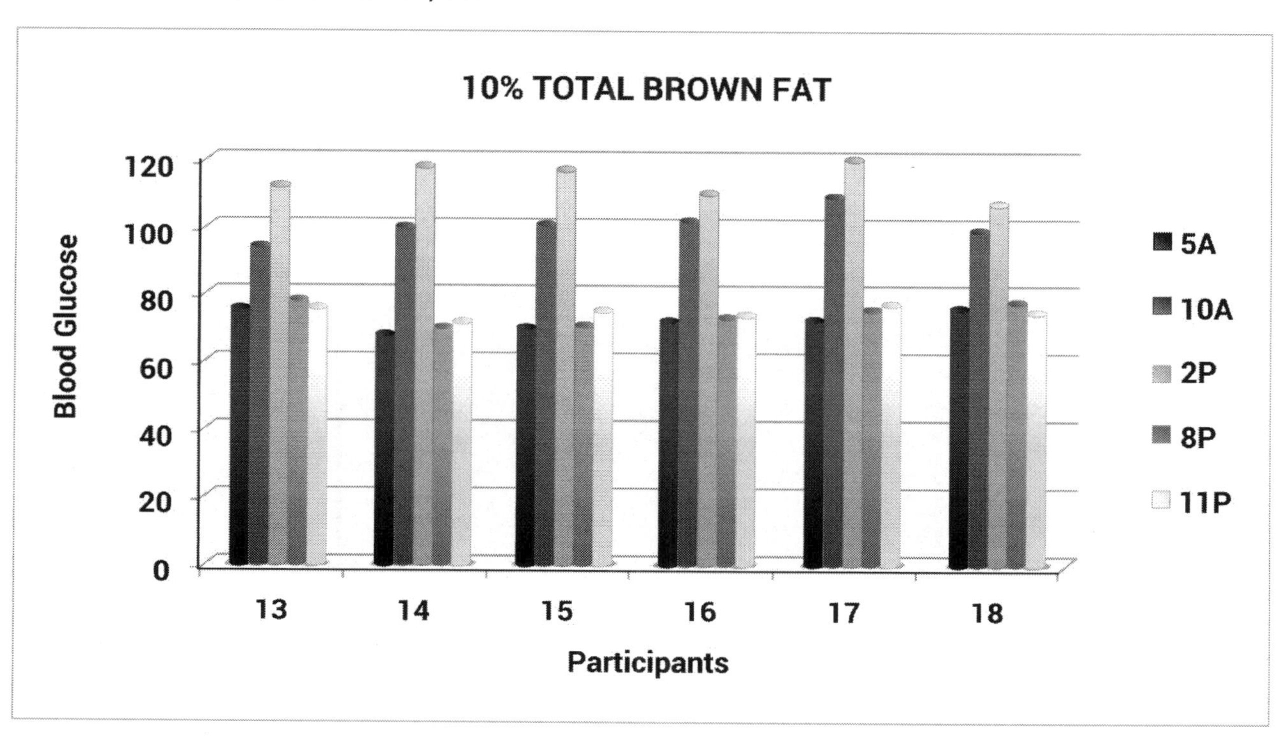

Figure 4 (below) identifies the resulting blood glucose measurements for participants with 20% total brown body fat.

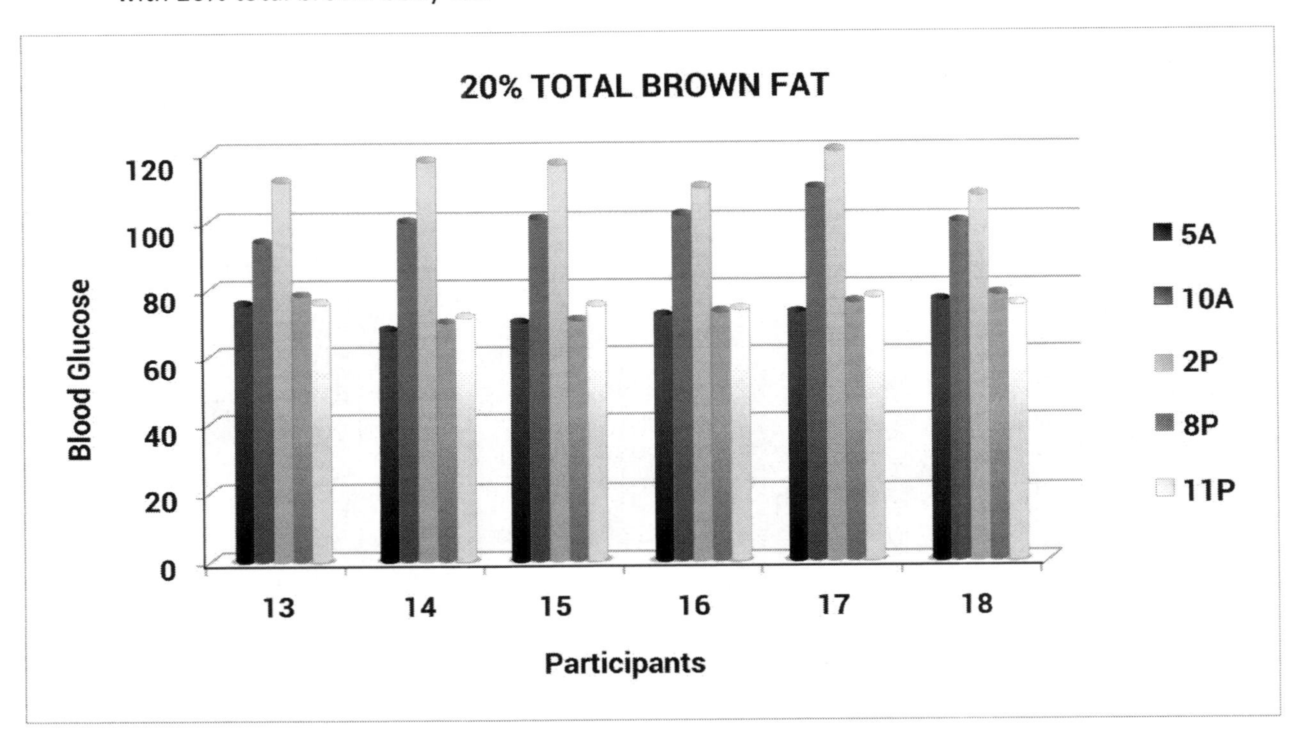

Figure 5 (below) identifies the average blood glucose measurements for the three trials.

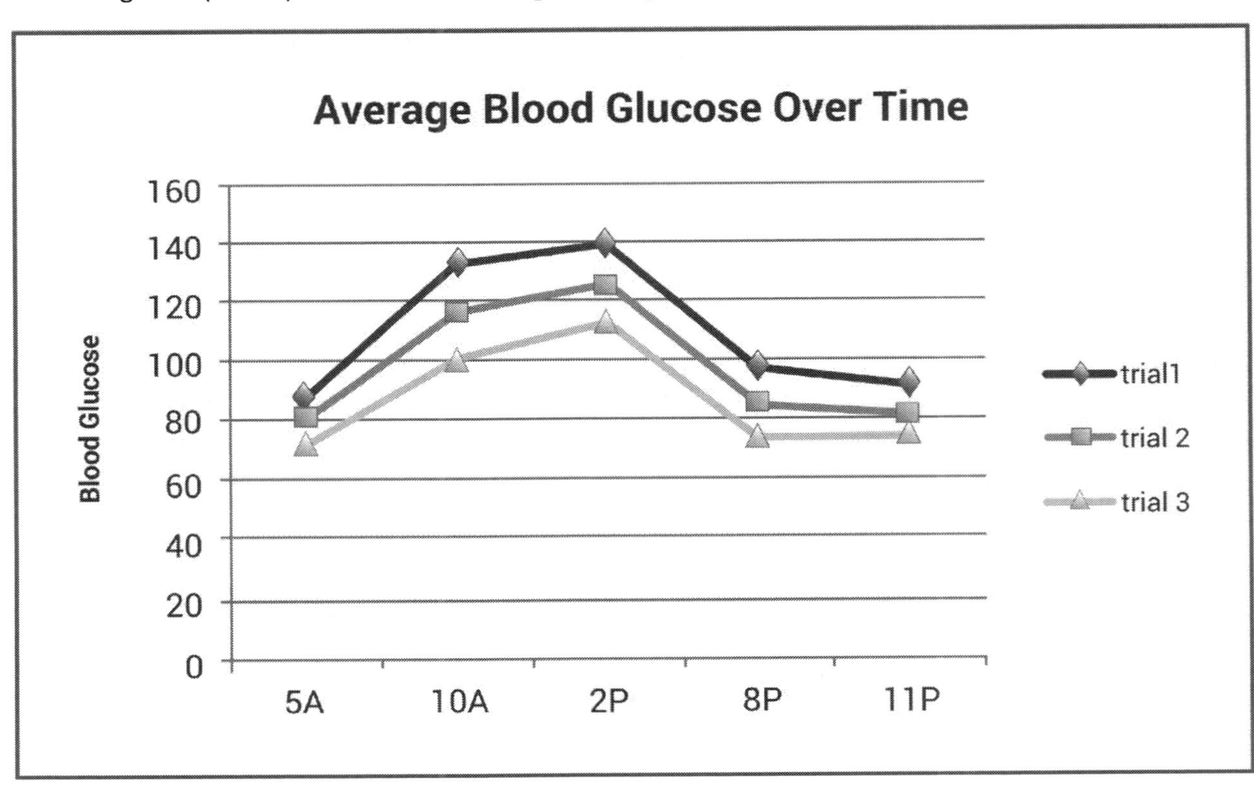

29. Which of the following describes the relationship of the research results in Figure 5?
    a. Positive correlation among the three trials
    b. Curvilinear relationship
    c. Weak negative relationship
    d. No demonstrated relationship

30. Which of the following statements concerning mitochondria is INCORRECT?
    a. Mitochondrial function is diminished in the presence of elevated blood glucose levels.
    b. Mitochondria are responsible for the color of brown fat.
    c. Mitochondria are capable of reproduction in response to energy needs.
    d. Mitochondria are responsible for binding oxygen in mature red blood cells.

31. According to Figure 5, participants' blood sugars were highest at what time of day?
    a. 5 a.m., because heat is generated early in the morning
    b. 8 p.m., because the participants ate less for dinner than lunch
    c. 11 p.m., because brown adipose tissue is not active at night
    d. 2 p.m., because the effects of the early-morning activity of the brown adipose tissue had diminished

32. Circadian rhythms control sleep by doing which of the following?
    a. Stimulating the pineal gland to release ganglia
    b. Stimulating the release of melatonin
    c. Suppressing shivering during cold mornings
    d. Instructing brown adipose tissue to release sugar

33. Which Participant in trial 1 had the highest average blood sugar for the group?
    a. 1
    b. 3
    c. 4
    d. 6

34. Is the data in Figure 5 consistent with the daytime plasma glucose trend in Figure 1?
    a. Yes, Figure 5 blood glucose readings declined from a morning to afternoon.
    b. No, blood glucose readings peaked at 2 p.m..
    c. Yes, morning glucose readings were higher in group 1.
    d. No, nighttime levels fluctuated between 100 and 110.

Passage 7

Questions 35-40 pertain to the following passage:

> A biome is a major terrestrial or aquatic environment that supports diverse life forms. Freshwater biomes—including lakes, streams and rivers, and wetlands—account for 0.01% of the Earth's fresh water. Collectively, they are home to 6% of all recognized species. Standing water bodies may vary in size from small ponds to the Great Lakes. Plant life in lakes is specific to the zone of the lake that provides the optimal habitat for a specific species, based on the depth of the water as it relates to light. The photic layer is the shallower layer where light is available for photosynthesis. The aphotic layer is deeper, and the levels of sunlight are too low for photosynthesis. The benthic layer is the bottom-most layer, and its inhabitants are nourished by materials from the photic

layer. Light-sensitive cyanobacteria and microscopic algae are two forms of phytoplankton that exist in lakes. As a result of nitrogen and phosphorous from agriculture and sewage run-off, algae residing near the surface can multiply abnormally so that available light is diminished to other species. Oxygen supplies may also be reduced when large numbers of algae die.

Recently, concerns have been raised about the effects of agriculture and commercial development on the quality of national freshwater bodies. In order to estimate the effect of human impact on freshwater, researchers examined plant life from the aphotic layer of three freshwater lakes of approximately the same size located in three different environments. Lake A was located in a remote forested area of western Montana. Lake B was located in central Kansas. Lake C was located in a medium-size city on the west coast of Florida. The researchers hypothesized that the microscopic algae and cyanobacteria populations from Lake A would approach appropriate levels for the size of the lake. They also hypothesized that the remaining two samples would reveal abnormal levels of the phytoplankton. In addition, the researchers measured the concentration of algae at different depths at four different times in another lake identified as having abnormal algae growth. These measurements attempted to identify the point at which light absorption in the photic layer was no longer sufficient for the growth of organisms in the aphotic layer. Resulting data is identified below.

Figure 1 (below) illustrates the zones of the freshwater lake.

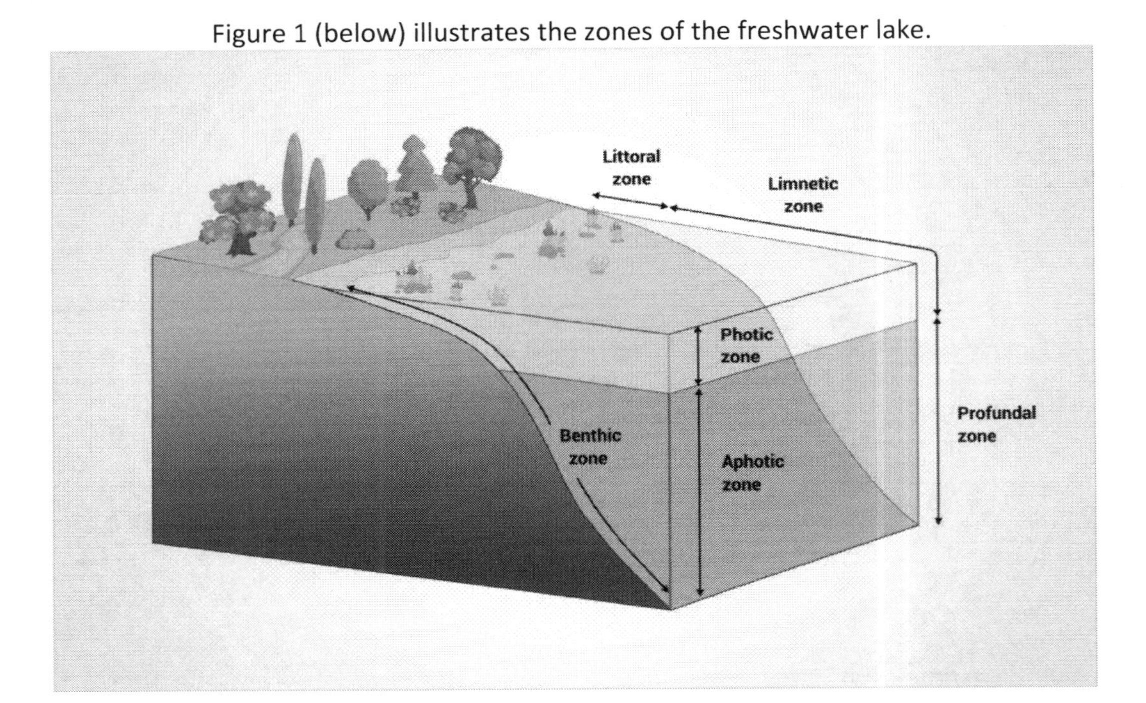

Figure 2 (below) identifies algae and cyanobacteria levels in parts per million for Lake A over six measurements.

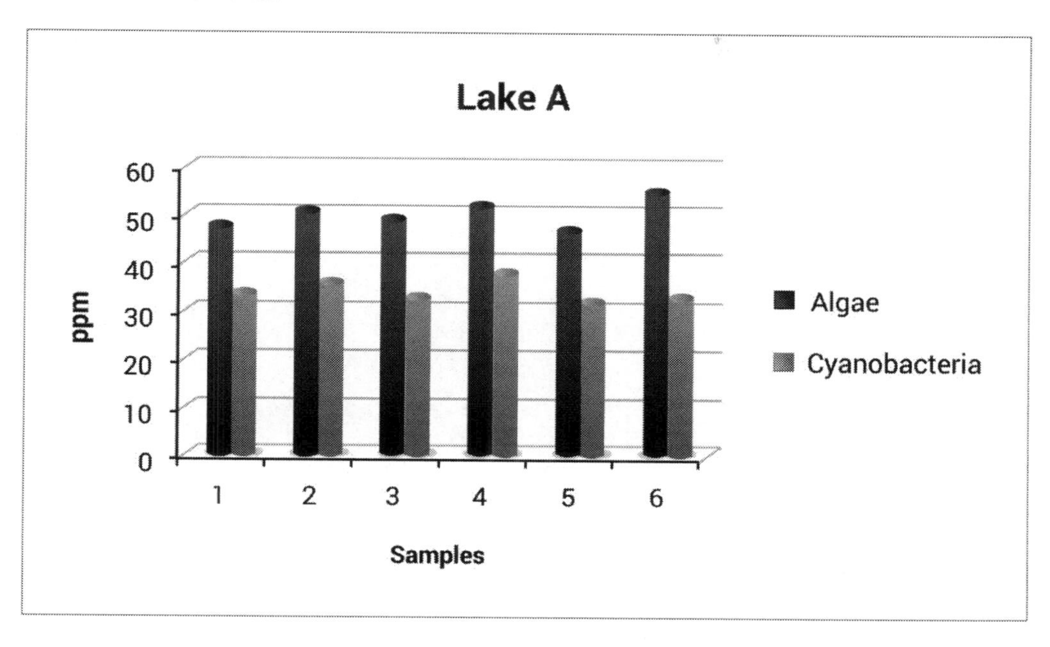

normal: Algae 50 p.p.m. Cyanobacteria 35 p.p.m.

Figure 3 (below) identifies algae and cyanobacteria levels in parts per million for Lake B over six measurements.

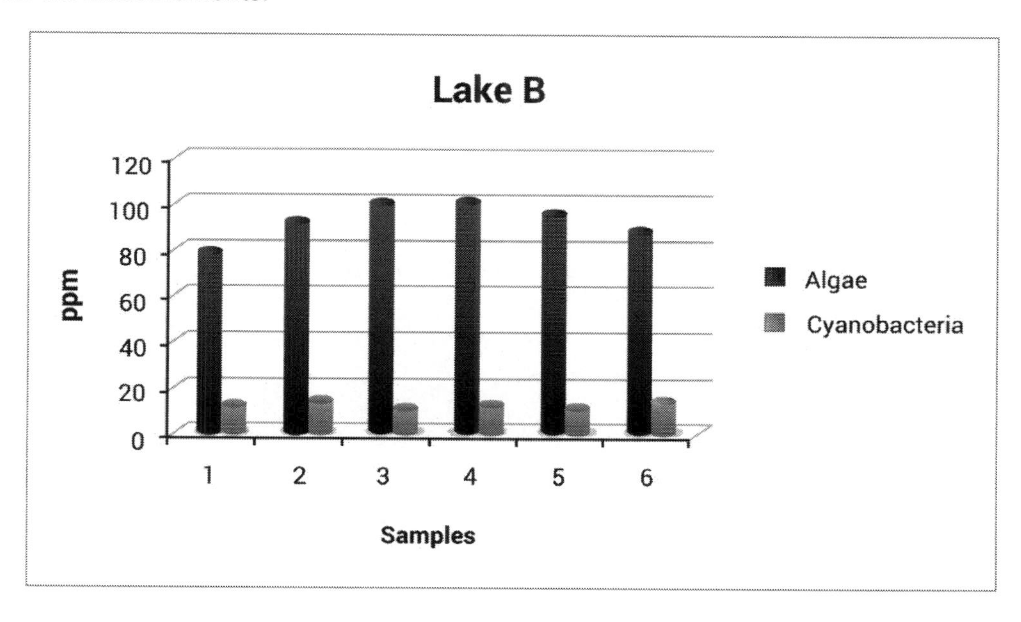

normal: Algae 50 p.p.m. Cyanobacteria 35 p.p.m.

Figure 4 (below) identifies algae and cyanobacteria levels in parts per million for Lake C over six measurements.

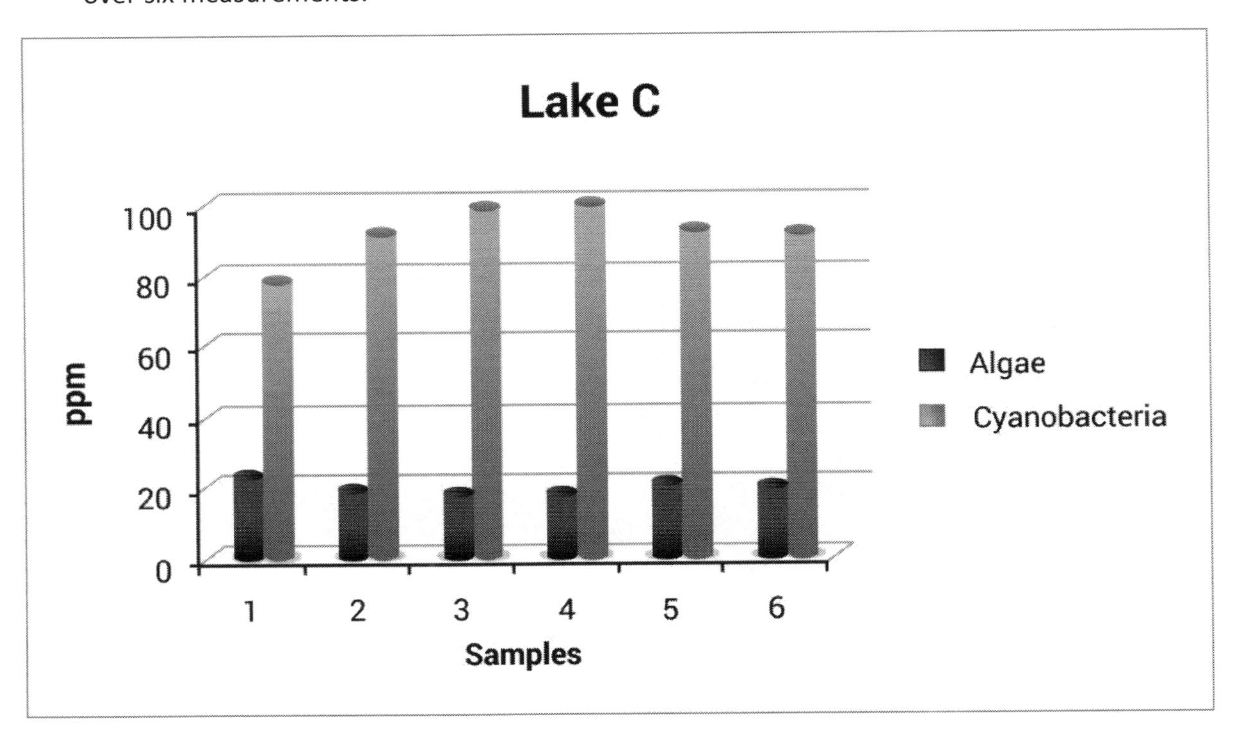

normal: Algae 50 p.p.m. Cyanobacteria 35 p.p.m.

Figure 5 (below) identifies cyanobacteria levels at different depths over time.

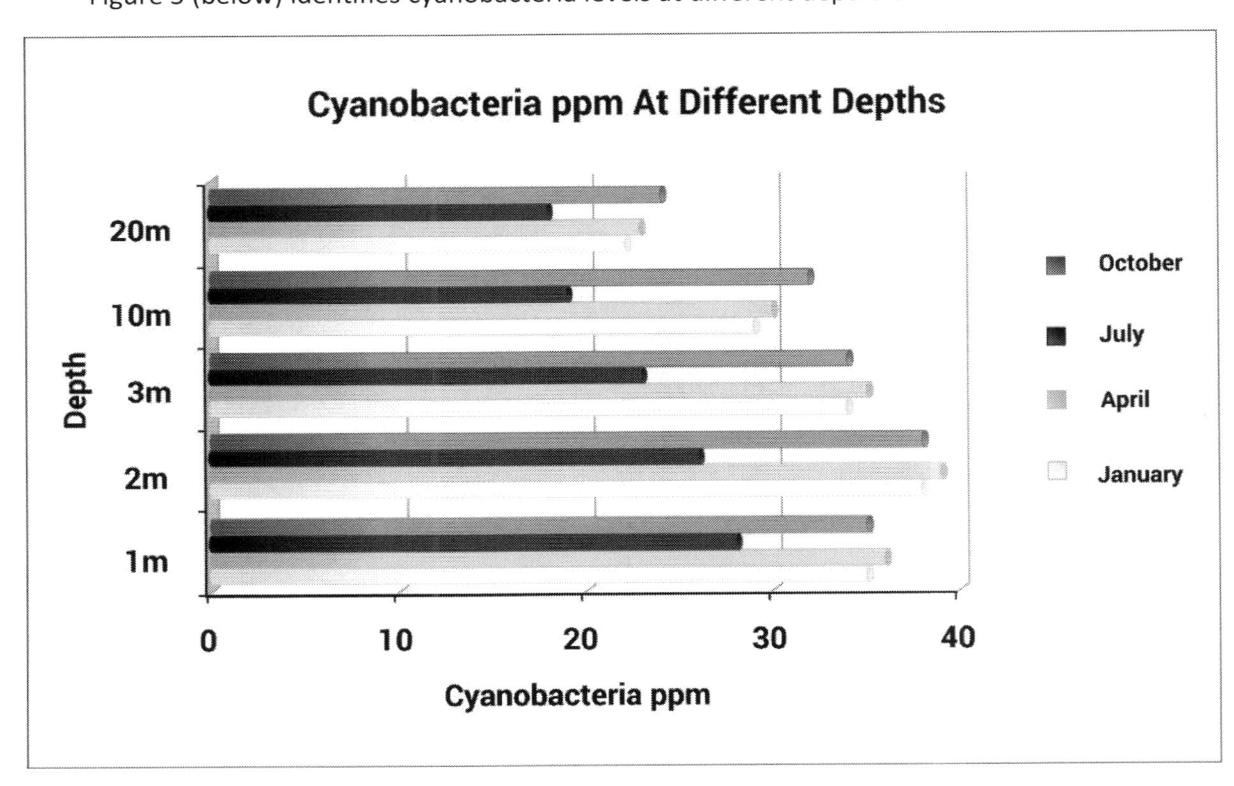

35. Based on Figure 2, was the researchers' hypothesis confirmed?
   a. No, the phytoplankton levels were not elevated in the first trial.
   b. Yes, the phytoplankton levels were raised above normal in each sample.
   c. No, the Lake A numbers were normal.
   d. Yes, algae levels were above normal in Lake C.

36. In Lake B, cyanobacteria were decreased and algae were increased. Which of the following is a possible explanation for this finding?
   a. The overgrowth of algae decreased the light energy available for cyanobacteria growth.
   b. Lake B experienced severe flooding, causing the water levels in the lake to rise above normal.
   c. Agricultural chemical residue depleted the food source for cyanobacteria.
   d. Cyanobacteria cannot survive in the cold winter weather in Lake B.

37. What common factor might explain the results for Lake B and Lake C?
   a. Population concentration
   b. Average humidity of the locations
   c. Average heat index
   d. Excess nitrogen and phosphorous in the ground water

38. As algae levels increase above normal, what happens to organisms in the aphotic level?
   a. Growth is limited but sustained.
   b. Species eventually die due to decreased oxygenation.
   c. Cyanobacteria increase to unsafe levels.
   d. Aerobic bacteria multiply.

39. Referencing Figures 2 and 3, which environment would favor organisms in the benthic layer of the corresponding lake?
   a. Figure 4, because the cyanobacteria are protective.
   b. Figure 3, because increased numbers of Algae provide more light.
   c. Figure 4, because cyanobacteria are able to survive.
   d. Figure 3, because the levels of both species are normal.

40. Which of the following statements is supported by the data in Figure 5?
   a. Algae growth is greater in July than April.
   b. Cyanobacteria can't exist at 20 meters in this lake.
   c. There's insufficient light in the aphotic layer at 3 meters to support algae growth.
   d. Cyanobacteria growth rates are independent of algae growth at 1 meter.

Answer Explanations #3

1. C: There is a 99% probability of PCR testing identifying *Histoplasma*. GM assay was more specific for identifying *Aspergillus,* 95% to 85%. True positive is defined by sensitivity. The sensitivity of GM assay testing is less than 70%.

2. D: *Histoplasma* is detectable 90 days from exposure. PCR testing is able to detect *Histoplasma* 91 days from exposure—one day after sufficient organisms exist for detection. *Candida* is detectable 45 days from exposure. PCR testing is able to detect *Candida* 72 days from exposure—27 days after a sufficient number of organisms exist for detection. *Aspergillus* is detectable 118 days from exposure. ELISA testing is able to detect *Aspergillus* 134 days from exposure—16 days after a sufficient number of organisms exist for detection. *Candida* is detectable 45 days from exposure. GM assay testing is able to detect *Candida* 56 days from exposure—11 days after a sufficient number of organisms exist for detection.

3. B: The probability that the GM assay will identify *Candida* is 69%. Therefore, there's a 31% probability that it won't be identified. ELISA sensitivity and specificity for *Histoplasma* are both greater than 80%. False-negative probabilities are represented by the specificity of a given testing method. The sensitivity and specificity for GM assay testing for *Aspergillus* is 9% and 96% respectively. All testing methods had greater than 90% specificity for the organisms.

4. C: The sensitivity of PCR testing for *Histoplasma* is 99%, and the test can identify the organism one day after it reaches a detectable colony size. The sensitivity for GM assay testing for *Histoplasma* is 65%. If physicians rely on GM assay testing, they may determine that the individual doesn't have the *Histoplasma* infection. Treatment will depend on the presence or absence of the infection as indicated by testing. Waiting for PCR testing is based on the sensitivity of the test, not the individual's current symptoms. The subclinical phase of *Histoplasma* is 28 days.

5. A: ELISA testing detects *Candida* three days after the organism is present in sufficient numbers to be recognized. PCR detects the organism more than three weeks after it is first detectable. ELISA testing sensitivity for *Candida* is 87% and PCR testing is 92%. However, the ability to identify the presence of the organism earlier in the process of infection (allowing early intervention) outweighs the differences in the probability of identifying the presence of the organism. There's a 92% probability that PCR testing will identify the presence of *Candida*. PCR testing is more sensitive than ELISA: 92% versus 87%

6. D: The main difference in the scientists' opinions is related to the cause of mad cow disease. The existence of species-specific proteins was used by Scientist 2 to support viral infection as the cause of the disease. Transmission rates of the disease and the conversion of normal proteins to prions were not debated in the passage.

7. A: Scientist 2 proposed that viruses were the cause of mad cow disease because chemicals inactivated the viruses. The remaining choices are correct.

8. B: According to Scientist 1, abnormal prions are capable of "refolding" normal proteins in harmful prions. Abnormal proteins accumulate to produce the damaging conglomerations. Scientist 2 didn't find species-specific DNA and used this fact to support viruses as the cause of mad cow disease. According to Scientist 1, prions are located in the central nervous system, not the peripheral nervous system.

9. D: Mad cow disease can be spread between animal species and from animals to humans through consumption of diseased animal products. The resulting damage to the central nervous system is

irreversible and will eventually cause the death of the animal. Scientist 2 would not agree that the infecting agent contained amino acids, as they form proteins, and Scientist 2 believes that a virus causes the disease. Scientist 2 demonstrated that the infected tissue of animals that were infected by a different species didn't contain species-specific DNA, which would have been the expected outcome if the infecting agent were a protein.

10. C: The accumulated masses of abnormal prions eventually form sponge-like holes in the brain and spinal cord that result in death. The passage doesn't mention the effects of the synapses, nerves, or blood supply.

11. D: The absence of disease resulting from the inactivated viral particles best supports the views of Scientist 2. There were no species-specific DNA sequences found in the infected particles. Scientist 2 didn't support the existence of prions as the cause of mad cow disease.

12. C: The actual process of "refolding" the normal protein into the abnormal protein isn't clear from this passage. Scientist 1 claims that prions cause the disease. Prions are an abnormal protein, not a virus. Scientist 1 claims that mad cow disease is caused by abnormal proteins.

13. C: The main hypothesis for this study involved the influence of 2-AG levels combined with sleep deprivation on eating behaviors. The combination of the two conditions, not each one separately, constitutes the main hypothesis. The passage didn't discuss a placebo effect in the normal saline injection group.

14. D: The study results support the hypothesis because the participants who received 1-AG injections and were sleep deprived gained more weight than participants who received sterile normal saline injections. The remaining choices do not support the hypothesis.

15. A: Participant H gained more than 1 pound (450g) per day. There was little fluctuation in the day-to-day weight gain for each participant. Participant H in trial 2 gained more weight than participant D in trial 1.

16. B: Sleep deprivation increases the levels and duration of action of 2-AG, an endogenous cannabinoid, especially in the late afternoon. The stress effect increases with the degree of sleep deprivation. The passage doesn't discuss a relationship between sleep deprivation and the hunger response. Eating behaviors are increased in late afternoon as a result of the extended duration of 2-AG action.

17. A: Circadian fluctuations increase the levels of 2-AG during the afternoon and evening. This increase is believed to stimulate food intake beyond the point of satiety. Endogenous cannabinoids decrease gastric motility. 2-AG may have a calming effect on mood, but food intake is still increased in the presence of afternoon and evening levels of 2-AG. Endogenous cannabinoids work with the opioid system to mediate the pain response, not food-seeking behaviors.

18. A: In trial 3, the plants grown with the combined-wavelength LED's reached 150 millimeters by day 21. The plants grown with white light reached 160 millimeters by day 35.

19. C: In trial 3, with LED lighting that included green and yellow wavelengths, plant growth was greater than trial 1 or trial 2 with either blue or red wavelengths. However, from the available information, it can only be said that green and yellow wavelengths *contributed to* plant growth in trial 3, but not that green and yellow wavelengths *alone* were responsible for plant growth in trial 3. There was plant growth in all lighting conditions.

20. B: In trial 1, from day 28 to day 35, white light growth increased by 71 millimeters, and red light increased by 78 millimeters. In trial 2, from day 28 to day 35, white light growth increased by 71 millimeters, and blue light increased by 78 millimeters.

21. C: The average daily growth with LED lighting was not twice the white light average daily growth. LED systems did result in better growth rates and they do require less water and electricity. However, the question is based on recorded average daily growth, and that rate was not double the white light rate.

22. D: The passage says that green and yellow wavelengths are reflected by the plant. Therefore, it's expected that those wavelengths would result in slower growth than the blue or red wavelengths, which are absorbed.

23. B: *Anacardiaceae Magnifera* is the genus and family name for the mango. The *Colletotrichum* fungus causes the spoilage. The remaining choices are correct.

24. C: According to Table 1, at 20 days, the fungal level at 7.5 °C was the same as the fungal level at 5 °C. Because the 7.5 °C temperature is less expensive than the 5 °C temperature, the 7.5°C model is best. The 10°C model is less expensive than the 7.5 °C, but fungal levels are greater. Only the 7.5 °C and 5 °C models had acceptable fungal levels at 20 days.

25. B: The hot water wash pre-treatment fungal level increased by 15 millimeters from day 26 to day 28 at 10 °C. It was the single largest one-day increase across the trials. Air cooling at 10 °C increased by 5 millimeters from day 14 to day 16. Air cooling at 7.5 °C increased by 7 millimeters from day 26 to day 28. Hot water wash at 7.5 °C increased 11 millimeters from day 26 to day 28.

26. C: The hot water wash fungal level at 10 °C reached 35 millimeters on day 20. The maximum fungal level for air cooling at 5 °C was 19 millimeters, and at 10 °C, 35 millimeters on day 26. The maximum fungal level for hot water at 7.5 °C was 36 millimeters on day 28.

27. C: The fungal levels were acceptable with the hot water wash at 7.5 °C, and the 7.5 °C temperature is less expensive to maintain. Air cooling and hot water wash pre-treatment at 5 °C resulted in acceptable fungal levels, but the 5 °C temperature is costlier. Fungal levels were not acceptable at 28 days at 10 °C.

28. B: Shipping mangoes at 5 °C is costly, but for the 28-day trip, the fungal levels were only acceptable in the 5 °C model. Air cooling fungal rates at 5 °C were lower than the hot water wash rates, but each was acceptable. Fungal rates at 7.5 °C and 10 °C were unacceptable.

29. A: The correlation was positive, because when one variable increased, the other increased, and when one variable decreased, the other decreased. There are two forms for a curvilinear relationship. In one curvilinear relationship, when variable 1 increases, a second variable increases as well, but only to a certain point, and then variable 2 decreases as variable 1 continues to increase. In the other form, variable 1 increases while variable 2 decreases to a certain point, after which both variables increase. In a negative relationship, high values for one variable are associated with low values for the second variable.

30. D: Mitochondrial activity is suppressed by elevated blood glucose levels. Mitochondria use sugar to produce cellular energy, and the presence of large numbers of mitochondria in BAT gives BAT a brownish color. Mitochondria contain DNA and can reproduce additional mitochondria when additional energy is required. In the body, mature red blood cells are the only cells that don't contain mitochondria.

31. D: Blood sugars for all groups identified in Figure 5 were highest at 2 p.m.

32. B: Circadian rhythms control sleep by stimulating the release of melatonin from the pineal gland. Ganglia are nerve cells, not hormones, that affect sleep. BAT doesn't release sugar; it utilizes sugar for heat production. Shivering on cold mornings is a desirable form of thermogenesis but isn't associated with sleep.

33. D: The average blood sugar for participant 6 was 115. Participant 1 was 105, participant 3 was 112, and participant 4 was 82.

34. B: The daytime blood glucose levels in Figure 1 decreased as the day progressed. The blood glucose levels in Figure 5 peaked for the day at 2 p.m. Night blood glucose levels didn't reach 100. Group I's levels are irrelevant to the question.

35. A: Based only on Figure 2, the researchers' hypothesis wasn't confirmed. Subsequent trials confirmed the hypothesis.

36. A: Increased algae levels can block sunlight, limiting growth of species inhabiting lower zones. The passage doesn't identify the effects of rainfall or cold temperatures on phytoplankton growth, so Choices *B* and *D* are incorrect. The passage identifies the effect of phosphorous and nitrogen residue on algae growth, but not as a food source for cyanobacteria.

37. D: The passage identifies freshwater contamination by phosphorous and nitrogen as the most common cause of algae overgrowth. Population density would be more common in Florida than Kansas.

38. B: As algae levels increase above normal, organisms in the aphotic level plants don't receive adequate light for normal growth and oxygen levels are decreased, resulting in the death of oxygen-dependent species.

39. C: Algae block the sunlight, which limits growth.

40. A: Algae growth was greater in July, which limited the amount of light reaching the lower zones of the lake, decreasing the levels of cyanobacteria. Cyanobacteria existed in less-than-normal concentrations at 20 meters, but there were measurable levels of the organisms. Algae growth at 3 meters wasn't measured. The passage states that cyanobacteria growth is associated with algae growth, not independent of algae growth.

Dear ACT Test Taker,

We would like to start by thanking you for purchasing this study guide for the science section of the ACT exam. We hope that we exceeded your expectations.

Our goal in creating this study guide was to cover all of the topics that you will see on the science section of the test. We also strove to make our practice questions as similar as possible to what you will encounter on test day. With that being said, if you found something that you feel was not up to your standards, please send us an email and let us know.

We would also like to let you know about other books in our catalog that may interest you.

Test Name	Amazon Link
ACCUPLACER	amazon.com/dp/1628456515
AP Biology	amazon.com/dp/1628456221
SAT	amazon.com/dp/1628456396
TSI	amazon.com/dp/162845511X

We have study guides in a wide variety of fields. If the one you are looking for isn't listed above, then try searching for it on Amazon or send us an email.

Thanks Again and Happy Testing!
Product Development Team
info@studyguideteam.com

Interested in buying more than 10 copies of our product? Contact us about bulk discounts:

bulkorders@studyguideteam.com

FREE Test Taking Tips DVD Offer

To help us better serve you, we have developed a Test Taking Tips DVD that we would like to give you for FREE. **This DVD covers world-class test taking tips that you can use to be even more successful when you are taking your test.**

All that we ask is that you email us your feedback about your study guide. Please let us know what you thought about it – whether that is good, bad or indifferent.

To get your **FREE Test Taking Tips DVD**, email freedvd@studyguideteam.com with "FREE DVD" in the subject line and the following information in the body of the email:

a. The title of your study guide.

b. Your product rating on a scale of 1-5, with 5 being the highest rating.

c. Your feedback about the study guide. What did you think of it?

d. Your full name and shipping address to send your free DVD.

If you have any questions or concerns, please don't hesitate to contact us at freedvd@studyguideteam.com.

Thanks again!

Made in the USA
San Bernardino, CA
13 January 2020

63129273R00095